P9-AEU-924

Extreme Value Distributions

Theory and Applications

Extreme Value Distributions

Theory and Applications

Samuel Kotz
Department of Engineering Management and Systems Engineering
The George Washington University, Washington DC, USA

Saralees Nadarajah
School of Mathematical Sciences, The University of Nottingham, UK

ICP
Imperial College Press

Published by

Imperial College Press
57 Shelton Street
Covent Garden
London WC2H 9HE

Distributed by

World Scientific Publishing Co. Pte. Ltd.

P O Box 128, Farrer Road, Singapore 912805

USA office: Suite 1B, 1060 Main Street, River Edge, NJ 07661

UK office: 57 Shelton Street, Covent Garden, London WC2H 9HE

Library of Congress Cataloging-in-Publication Data

Kotz, Samuel.

Extreme value distributions : theory and applications / Samuel Kotz, Saralees Nadarajah.

p. cm.

Includes bibliographical references and index.

ISBN 1860942245 (alk. paper)

1. Extreme value theory. I. Nadarajah, Saralees.

QA273.6.K682 2000
519.2'4--dc21 00-039667

British Library Cataloguing-in-Publication Data

A catalogue record for this book is available from the British Library.

This book is printed on acid-free paper.

Printed in Singapore by Uto-Print

Preface

This monograph attempts to describe in an organized manner the central ideas and results of probabilistic extreme-value theory and related models stemming from the pioneering contributions of E. J. Gumbel in the early forties of this century. The exposition is unencumbered by excessive mathematical details and almost no proofs are provided. It is a book *about* extreme-value distributions — both univariate and multivariate — and their applications, supplemented by an up-to-date extensive bibliography, aimed mainly at a novice in the field; hopefully a specialist may find therein some useful information as well.

By laying bare the main structure of the theory of extreme value distributions and its applications, including the assumptions and conclusions, deficiencies and advantages, it is our intention that the volume will serve as a useful, balanced and critical introduction and simultaneously a guide to the literature. We have tried to keep the language and notation sufficiently familiar and simple to make it accessible for scientists with a modest probabilistic background. As always, as it is the case for books on probability, statistics and in particular on distribution theory, the ill-defined quality of "mathematical sophistication" and the ability to connect empirical statements with rigorous mathematical deductions are desirable prerequisites. In our opinion, the extreme value theory — as described in this book — is a most important and successful example of applicability of mathematics to modern engineering, empirical and environmental problems of great significance, and it is our hope that we also were somewhat successful in conveying the message.

The authors express their thanks to Professors N. D. Singpurwalla and J. A. Tawn for their useful comments and to the editors at the Imperial College Press for their most helpful assistance and efficient handling of the manuscript.

Samuel Kotz, Washington, D.C., U.S.A.
Saralees Nadarajah, Nottingham, U.K.

September, 1999

Contents

Chapter 1

Univariate Extreme Value Distributions

1.1 Historical Survey

Probabilistic extreme value theory is a curious and fascinating blend of an enormous variety of applications involving natural phenomena such as rainfall, floods, wind gusts, air pollution, and corrosion, and delicate advanced mathematical results on point processes and regularly varying functions. This area of research thus attracted initially the interests of theoretical probabilists as well as engineers and hydrologists, and only relatively recently of the mainstream statisticians. For a number of years it was closely related to the activities of E. J. Gumbel, a colorful personality, whose life and activities were affected by pre-World War II upheavals.

The following pages are addressed not only or primarily to professionals in the field of statistical distributions and statistical inference but to that much larger audience which is interested in the topics without willing or being able to devote more than a limited amount of time to considering them.

Probabilistic extreme value theory, first of all, deals with the stochastic behaviour of the maximum and the minimum of i.i.d. random variables. The distributional properties of extremes (maximum and minimum), extreme and intermediate order statistics, and exceedances over (below) high (low) thresholds are determined by the upper and lower tails of the underlying distribution.

Conversely, the tail of the underlying distribution function or functional parameters thereof may be evaluated by means of statistical procedures based on extreme and intermediate order statistics or exceedances over high thresholds. Focussing our attention on the tails has the advantage that certain parametric statistical models, specifically tailored for that part of the distribution, can be introduced.

Historically, work on extreme value problems may be traced back to as early as 1709 when Nicolas Bernouilli discussed the mean largest distance from the origin given n points lying at random on a straight line of a fixed length t [see Gumbel (1958)].

Extreme value theory has originated mainly from the needs of astronomers in utilizing or rejecting outlying observations. The early papers by Fuller (1914) and Griffith (1920) on the subject were specialized both in fields of applications and in methods of mathematical analysis (see below). A systematic development of the general theory may be regarded as having started with the paper by von Bortkiewicz (1922) that dealt with the distribution of range in random samples from a normal distribution. The importance of the paper by Bortkiewicz is inherent by the fact that the concept of *distribution of largest value* was clearly introduced in it for the first time. (In his classical book, E. J. Gumbel (1958) devotes a chapter to the memory of L. von Bortkiewicz.) In the very next year von Mises (1923) evaluated the expected value of this distribution, and Dodd (1923) calculated its median, also discussing some nonnormal parent distributions. Of more direct relevance is a paper by Fréchet (1927) in which asymptotic distributions of largest values are considered. In the following year Fisher and Tippett (1928) published results of an independent inquiry into the same problem. While Fréchet (1927) had identified one possible limit distribution for the largest order statistic, Fisher and Tippett (1928) showed that extreme limit distributions can *only* be one of three types. Tippett (1925) had earlier studied the exact cumulative distribution function and moments of the largest order statistic and of the sample range from a normal population. Von Mises (1936) presented some simple and useful *sufficient* conditions for the weak convergence of the largest order statistic to each of the three types of limit distributions given earlier by Fisher and Tippett (1928). We shall discuss von Mises' conditions in a subsequent section. In 1943, Gnedenko presented a rigorous foundation for the extreme value theory and provided necessary and sufficient conditions for the weak convergence of the extreme order statistics.

Mejzler (1949), Marcus and Pinsky (1969) (unaware of Mejzler's result) and de Haan (1970) (1971) refined the work of Gnedenko. An important but much neglected work of Juncosa (1949) extends Gnedenko's results to the case of not necessarily identically distributed independent random variables. Although of strong theoretical interest, Juncosa's results do not seem to have much practical utility. The fact that asymptotic distributions of a very general nature can occur does not furnish much guidance for practical applications.

The theoretical developments of the 1920s and mid 1930s were followed in the late 1930s and 1940s by a number of papers dealing with practical applications of extreme value statistics in distributions of human lifetimes, radioactive emissions [Gumbel (1937a,b), strength of materials [Weibull (1939)], flood analysis [Gumbel (1941, 1944, 1945, 1949a), Rantz and Riggs (1949)], seismic analysis [Nordquist (1945)], and rainfall analysis [Potter (1949)] to mention a few examples. From the application point of view, Gumbel made several significant contributions to the extreme value analysis; most of them are detailed in his book-length account of statistics of extremes [Gumbel (1958)]. See the sections on Applications for more details.

Gumbel was the first to call the attention of engineers and statisticians to possible applications of the formal "extreme-value" theory to certain distributions which had previously been treated empirically. The first type of problem treated in this manner in the USA had to do with meteorological phenomena — annual flood flows, precipitation maxima, etc. This occurred in 1941.

In essence, all the statistical models proposed in the study of fracture take as a starting point Griffith's theory (already alluded to above), which states that the difference between the calculated strengths of materials and those actually observed resides in the fact that there exist flaws in the body which weaken it.

The first writer to realize the connection between specimen strength and distribution of extreme values seems to be F. T. Peirce (1926) of the British Cotton Industry Association. The application of essentially the same ideas to the study of the strength of materials was carried out by the well-known Swedish physicist and engineer, W. Weibull (1939).

The Russian physicists, Frenkel and Kontorova (1943), were the next to study these problems. Another important neglected early publication related to extreme value analysis of the distribution of feasible strengths of rubbers is due to S. Kase (1953).

A comprehensive bibliography of literature on extreme value distributions and their applications can easily be constructed to contain over 1,000 items at the time of this writing (1999). While this extensive literature serves as a testimony to the great vitality and applicability of the extreme value distributions and processes, it also unfortunately reflects on the lack of coordination between researchers and the inevitable duplication (or even triplication) of results appearing in a wide range of diverse publications.

There are several excellent books that deal with the asymptotic theory of extremes and their statistical applications. We cite a few known to us (without in any way dispraising those that are not mentioned). David (1981) and Arnold, Balakrishnan, and Nagaraja (1992) provide a compact account of the asymptotic theory of extremes; Galambos (1978, 1987), Resnick (1987), and Leadbetter, Lindgren, and Rootzén (1983) present elaborate treatments of this topic. Reiss (1989) discussed various convergence concepts and rates of convergence associated with extremes (and also with order statistics). Castillo (1988) has successfully updated Gumbel (1958) and presented many statistical applications of extreme value theory with emphasis on engineering. Harter (1978) prepared an authoritative bibliography of extreme value theory which is still of substantial scientific value. Beirlant, Teugels and Vynekier (1996) provide a lucid practical analysis of extreme values with emphasis on actuarial applications.

1.2 The Three Types of Extreme Value Distributions

Extreme value distributions are usually considered to comprise the following three families:

Type 1, (Gumbel-type distribution):

$$\Pr[X \le x] = \exp[-e^{(x-\mu)/\sigma}]\,. \tag{1.1}$$

Type 2, (Fréchet-type distribution):

$$\Pr[X \le x] = \begin{cases} 0\,, & x < \mu\,, \\ \exp\left\{-\left(\dfrac{x-\mu}{\sigma}\right)^{-\xi}\right\}, & x \ge \mu\,. \end{cases} \tag{1.2}$$

Type 3, (Weibull-type distribution):

$$\Pr[X \le x] = \begin{cases} \exp\left\{-\left(\dfrac{\mu-x}{\sigma}\right)^{\xi}\right\}, & x \le \mu \\ 0 & x > \mu \end{cases} \tag{1.3}$$

where μ, $\sigma(>0)$ and $\xi(>0)$ are parameters.

The corresponding distributions of $(-X)$ are also called extreme value distributions. (Observe that Fréchet and Weibull distributions are related by a simple change of sign.)

Of these families of distributions, type 1 is the most commonly referred to in discussions of extreme values. Indeed some authors call (1.1) *the* extreme value distributions. In view of this, and the fact that distributions (1.2) and (1.3) can be transformed to type 1 distributions by the simple transformations

$$Z = \log(X - \mu), \qquad Z = -\log(\mu - X),$$

respectively, we will, for the greater part of this chapter, confine ourselves to discussion of type 1 distributions. We note that type 3 distribution of $(-X)$ is a *Weibull* distribution.

Type 1 distributions are also sometimes called, in earlier writings, doubly exponential distributions, on account of the functional form of (1.1).

The term "extreme value" is attached to these distributions because they can be obtained as limiting distributions (as $n \to \infty$) of the greatest value among n independent random variables each having the same continuous distribution. By replacing X by $-X$, limiting distributions of least values are obtained.

Although the distributions are known as extreme value, it is to be borne in mind that they do not represent distributions of all kinds of extreme values (e.g., in samples of finite size), and they can be used empirically (without an extreme value model).

The three types of distributions in (1.1)–(1.3) may all be represented as members of a single family of generalized distributions with cumulative distribution function

$$\Pr[X \le x] = \left[1 + \xi\left(\frac{x-\mu}{\sigma}\right)\right]^{-1/\xi}, \quad 1 + \xi\left(\frac{x-\mu}{\sigma}\right) > 0, \quad -\infty < \xi < \infty, \quad \sigma > 0. \tag{1.4}$$

For $\xi > 0$, the distribution (1.4) is of the same form as (1.2). For $\xi < 0$, the distribution (1.4) becomes of the same form as (1.3). Finally, when $\xi \to \infty$ or $-\infty$, the distribution (1.4) becomes the same form as the type 1 extreme value distribution in (1.1). For this reason the distribution function in (1.4) is known as the *generalized extreme value distribution* and is also sometimes referred to as the *von Mises type extreme value distribution* or the *von Mises–Jenkinson type distribution.* We shall discuss this generalized family in a separate section. Occasionally we shall use slightly different, but of course equivalent parametrizations.

1.3 Limiting Distributions and Domain of Attraction

Extreme value distributions were obtained as limiting distributions of greatest (or least) values in random samples of increasing size. To obtain a nondegenerate limiting distribution, it is necessary to "reduce" the actual greatest value by applying a linear transformation with coefficients which depend on the sample size. This process is analogous to standardization though not restricted to this particular sequence of linear transformations.

If $X_1, X_2, \ldots, X_n$ are independent random variables with common probability density function

$$p_{X_j}(x) = f(x), \qquad j = 1, 2, \ldots, n,$$

then the cumulative distribution function of $X'_n = \max(X_1, X_2, \ldots, X_n)$ is

$$F_{X'_n}(x) = [F(x)]^n,$$

where

$$F(x) = \int_{-\infty}^{x} f(x)dt.$$

As n tends to infinity, it is clear that for any fixed value of x

$$\lim_{n\to\infty} F_{X'_n}(x) = \begin{cases} 1 & \text{if } F(x) = 1\,, \\ 0 & \text{if } F(x) < 1\,, \end{cases}$$

which is a degenerate distribution. If there is a limiting distribution of interest, we must find it as the limiting distribution of some sequence of *transformed* "reduced" values, such as $(a_n X'_n + b_n)$ where a_n, b_n may depend on n but not on x.

To distinguish the limiting cumulative distribution of the "reduced" greatest value from $F(x)$, we will denote it by $G(x)$. Then since the greatest of Nn values $X_1, X_2, \ldots, X_{Nn}$ is also the greatest of the N values

$$\max(X_{(j-1)n+1}, X_{(j-1)n+2}, \ldots, X_{jn})\,, \qquad j = 1, 2, \ldots, N\,,$$

it follows that $G(x)$ must satisfy the equation

$$[G(x)]^N = G(a_N x + b_N)\,. \tag{1.5}$$

This equation was obtained by Fréchet (1927) and also by Fisher and Tippett (1928). It is sometimes called the *stability postulate.*

Type 1 distributions are obtained by taking $a_N = 1$; types 2 and 3 by taking $a_N \neq 1$. In this latter case

$$x = a_N x = b_N \qquad \text{if } x = b_N(1 - a_N)^{-1}\,,$$

and from (1.5) it follows that $G(b_N(1 - a_N)^{-1})$ must equal to 1 or 0. Type 2 corresponds to 1, and type 3 to 0. We shall briefly sketch the derivation of the type 1 distribution ($a_N = 1$). Equation (1.5) is now

$$[G(x)]^N = g(x + b_N)\,. \tag{1.6}$$

Since $G(x + b_N)$ must also satisfy (1.5)

$$[G(x)]^{NM} = [G(x + b_N)]^M = G(x + b_N + b_M)\,. \tag{1.7}$$

Also from (1.5)

$$[G(x)]^{NM} = G(x + b_{NM}) \tag{1.8}$$

and from (1.7) and (1.8) we have

$$b_N + b_M = b_{NM}\,,$$

whence

$$b_n = \sigma \log N\,, \qquad \text{with } \sigma \text{ a constant}\,. \tag{1.9}$$

Taking logarithms of (1.6) *twice* and inserting the value of b_N from (1.9), we have (noting that $G \leq 1$)

$$\log N + \log\{-\log G(x)\} = \log\{-\log G(x + \sigma \log N)\}\,. \tag{1.10}$$

Denote

$$h(x) = \log\{-\log G(x)\}\,.$$

Hence from (1.10) we have

$$h(x) = h(0) - \frac{x}{\sigma}\,. \tag{1.11}$$

Since $h(x)$ decreases as x increases, $\sigma > 0$. From (1.11),

$$\begin{aligned} -\log G(x) &= \exp\left[-\frac{x - \sigma h(0)}{\sigma}\right] \\ &= \exp\left(-\frac{x-\mu}{\sigma}\right), \end{aligned}$$

where $\mu = \sigma \log(-\log G(0))$. Hence

$$G(x) = \exp\left[-e^{-(x-\mu)/\sigma}\right],$$

in agreement with (1.1). For derivations of types 2 and 3, interested readers may refer to Galambos (1978, 1987).

As already mentioned earlier, Gnedenko (1943) established certain correspondences between the parent distribution $[F(x)]$ and the type to which the *limiting distribution* belongs. It should be noted that the conditions relate essentially to the behavior of $F(x)$ for high (low) values of x if the limiting distribution of greatest (least) values is to be considered. It is possible for greatest and least values, corresponding to the same parent distribution, to have different limiting distributions.

The conditions established by Gnedenko can be summarized as follows:

For the type 1 distribution: Defining X_α by the equation

$$F(X_\alpha) = \alpha\,,$$

the condition is

$$\lim_{n\to\infty} n[(1 - F(X_{1-n^{-1}} + y(X_{1-(ne)^{-1}} - X_{1-n^{-1}})))] = e^{-y}\,. \tag{1.12}$$

For the type 2 distribution:

$$\lim_{x\to\infty} \frac{1 - F(x)}{1 - F(cx)} = c^k\,, \qquad c > 0, k > 0\,. \tag{1.13}$$

For the type 3 distribution:

$$\lim_{x\to 0^-} \frac{1 - F(cx+\omega)}{1 - F(x+\omega)} = c^k\,, \qquad c > 0, k > 0\,, \tag{1.14}$$

where $F(\omega) = 1$, $F(x) < 1$ for $x < \omega$.

Gnedenko also showed that these conditions are necessary, as well as sufficient, and that there are no other distributions satisfying the stability postulate. Among distributions satisfying the type 1 condition (22.13) are normal, exponential and logistic; the type 2 condition (22.14) is satisfied by Cauchy; the type 3 condition is satisfied by nondegenerate distributions with range of variation bounded above.

Gnedenko's (1943) results have been generalized by numerous authors. Results for order statistics of fixed and increasing rank were obtained by Smirnov (1952) who — in his extremely valuable theoretical paper — completely characterized the limiting types and their domains of attraction. Generalizations for the maximum term have been made by Juncosa (1949) who dropped the assumption of a common distribution, Watson (1954) who proved that under mild restrictions the limiting distribution of the maximum term of a stationary sequence of m-dependent random variables is the same as in the independent case, Berman (1962) who studied exchangeable random variables and samples of random size, and Harris (1970) who extended the classical theory by introducing a model from reliability theory. Weinstein (1973) generalized the basic result of Gnedenko dealing with the asymptotic distribution of the exponential case with the initial distribution

$$V(x) = 1 - e^{-x^{\nu}} \qquad (x \geq 0)$$

[Gnedenko's (1943) result is for $\nu = 1$]. Jeruchim (1976) has warned that the additional parameter ν must be treated cautiously in applications.

Gnedenko's derivations resulted in further investigations as to validity of the law of large numbers for maxima and the relative stability of maxima. Green (1975) derived sufficient conditions for the consistent estimation of parameters based on extremes for very broad families of distributions.

There are numerous extensions of limit theorems for extremes in general settings. For a most recent discussion see Silvestrov and Teugels (1998)[a] and the references therein. The results require advanced tools such as Skorokhod topology J, U-compactness, etc. which are beyond the scope of this monograph. So far these results are solely of theoretical interest.

The necessary and sufficient conditions in (1.12)–(1.14) are often difficult to verify. In such instances the following sufficient conditions established by von Mises (1936) may be useful (though they are applicable only for absolutely continuous parent distributions):

As mentioned above, we start with underlying *common* distribution function F. Let $X_1, \ldots, X_n$ be i.i.d. random variables with the distribution F. Let $X_{1:n} \leq \cdots \leq X_{n:n}$ be the corresponding order statistics. Then:

$(X_{n:n} - b_n)/a_n$ converges in distribution to some nondegenerate limiting distribution G for some choice of constants $a_n > 0$, $b_n \in R$ *if and only if* for any integer k the random vector

$$((X_{n-i+1:n} - b_n)/a_n)_{i=1}^{k}$$

converges in distribution to $G^{(k)}$ having Lebesgue density

$$g^{(k)}(x_1, \ldots, x_k) = \begin{cases} G(x_k)\displaystyle\prod_{i=1}^{k} G'(x_i)/G(x_i) & \text{if } x_1 > \cdots > x_k, \\ 0 & \text{otherwise}. \end{cases}$$

In this case we say that F is in the *domain of attraction of* G and the notation $F \in D(G)$ is by now universally acceptable.

There is a multitude of papers in which necessary and sufficient conditions for F to be in the domain of attraction of G are provided.

Among these, the sufficient conditions due to von Mises (1936) are widely studied to this day and are easily applicable.

[a]Silvestrov, D. S. and Teugels, J. L. (1998), *Limit theorems for extremes with random sample size*, *Adv. Appl. Prob.* **30**, 777–800.

Recall that G must be of one of the following types, where $\alpha > 0$:

$$G_{1,\alpha}(x) := \exp\left(-x^{-\alpha}\right), \qquad x > 0\,, \qquad \text{(Fréchet)},$$
$$G_{2,\alpha}(x) := \exp\left(-(-x)^{\alpha}\right), \qquad x \le 0\,, \qquad \text{(reversed Weibull) and}$$
$$G_3(x) := \exp\left(-e^{-x}\right), \qquad x \in R\,, \qquad \text{(Gumbel distribution)}\,.$$

Assume that F possesses a *positive* derivative f on $[x_0, \omega(F)]$ where $x_0 < \omega(F)$ and $\omega(F)$ is $\sup\{x \in R$ for which $F(x) < 1\}$. Then, according to von Mises, in order that $F \in D(G_{1,\alpha})$ it is sufficient that $\omega(F) = \infty$ (i.e. for all $x \in R$, $F(x) < 1$), and $\lim_{x\to\infty} xf(x)/[1-F(x)] = \alpha$, with the very same α as in the definition of $G_{1,\alpha}$.

A more recent necessary and sufficient condition due to Galambos (1987) and Worms (1998) is based on the *additive excess property*:

There exists a mapping g from $]-\infty, s(P)[$ (where $s(P) = \sup\{x \in R; 1 - F(x) > 0\}$) into R_+ such that for all $x \in R_+$:

$$\lim_{t\to s(P)} [1 - F(t + g(t)x)]/[1 - F(t)] = e^{-x}.$$

It was Worms (1998) who replaced Galambos' (1987) condition $x \in R$ (which has no probabilistic meaning) by the condition $x \in R_+$.

For $F \in D(G_2, \alpha)$ it is sufficient that

$$\omega(F) < \infty \qquad \text{and} \qquad \lim_{x\uparrow\omega(F)} \frac{(\omega(F) - x)f(x)}{1 - F(x)} = \alpha\,,$$

(namely $F(x) < 1$ only on a finite set; recall the definition of $G_{2,\alpha}$).

For $F \in D(G_3)$ the von Mises condition is a bit more involved. In the original version it is:

$$\int_{-\infty}^{\omega(F)} (1 - F(u))du < \infty \qquad \text{and} \qquad \lim_{x\uparrow\omega(F)} f(x) \int_x^{\omega(F)} (1 - F(u))du/(1 - F(x))^2 = 1\,.$$

Falk and Marohn (1993) suggested the following strengthening of this condition: Suppose F has a positive derivative f on $(x_0, \omega(F))$ such that for some $c \in (0, \infty)$,

$$\lim_{x\to\omega(F)} \frac{f(x)}{1 - F(x)} = c\,.$$

Since $(1 - F(x))' = -f(x)$ and $\int_x^{\omega(F)}((1 - F(u))du)' = -(1 - F(x))$,

$$\lim_{x\to\omega(F)} \frac{f(x)}{1 - F(x)} = c \qquad \text{implies} \qquad \lim_{x\to\omega(F)} f(x) \int_x^{\omega(F)} \frac{(1 - F(u))}{(1 - F(x))^2} du = 1$$

by the l'Hôpital rule.

Consider the normal density φ and the corresponding distribution function Φ (*not* an extreme value distribution). For this density it is easy to verify that

$$\lim_{x\to\infty} \frac{x(1 - \Phi(x))}{\varphi(x)} = 1\,.$$

Thus Φ satisfies the original von Mises condition but not the Falk and Marohn (1993) modification.

On the other hand, there are examples of distributions for which the original, but not the modified condition, are satisfied. Falk and Marohn (1993) provided an alternative (equivalent) formulation of von Mises conditions via the generalized Pareto distributions (GPD) rather than the extreme value distributions. They take the distribution functions

$$W(x) = 1 + \log(G(x))$$

which yield the three classes of distributions:

$$W_{1,\alpha}(x) = 1 - x^{-\alpha}, \qquad x > 1$$
$$W_{2,\alpha}(x) = 1 - (-x)^{\alpha}, \qquad x \in [-1, 0]$$

and

$$W_3(x) = 1 - \exp(-x), \qquad x \geq 0.$$

Note that $G(x)$ must satisfy

$$\frac{1}{e} \leq G(x) \leq 1.$$

The choice of the normalizing constants a_N and $b_N > 0$ — to be denoted from now on by a_n and b_n — are as follows:

For the type 1 distribution:

$$\begin{aligned} a_n &= F^{-1}\left(1 - \frac{1}{n}\right), \\ b_n &= F^{-1}\left(1 - \frac{1}{ne}\right) - F^{-1}\left(1 - \frac{1}{n}\right). \end{aligned} \tag{1.15}$$

For the type 2 distribution:

$$\begin{aligned} a_n &= 0, \\ b_n &= F^{-1}\left(1 - \frac{1}{n}\right). \end{aligned} \tag{1.16}$$

For the type 3 distribution:

$$\begin{aligned} a_n &= F^{-1}(1), \\ b_n &= F^{-1}(1) - F^{-1}\left(1 - \frac{1}{n}\right). \end{aligned} \tag{1.17}$$

Analogous results for the limiting distributions of the sample minimum can be stated in a straightforward manner. Another line of development is the characterization of convergence in terms of moments. Clough and Kotz (1965) suggested using the mean and standard deviation of the distribution as scaling constants in place of b_n and a_n. This is valid, provided the rescaled mean and variance themselves converge to the mean and variance of the limiting distribution. Conditions for that were given by Pickands (1968).

With $F_X(x;\mu,\sigma)$ denoting the extreme value distribution for the sample minimum with the cdf given by

$$F_X(x;\mu,\sigma) = 1 - \exp\{-e^{(x-\mu)/\sigma}\}, \qquad \sigma > 0, \qquad \mu \in R,$$

and $G_X(x;a,b,c)$ denoting the three-parameter Weibull distribution with the cdf

$$G_X(x;a,b,c) = \begin{cases} 0 & \text{if } x \le c, \\ 1 - e^{-[(x-c)/b]^a} & \text{if } x \ge c, \end{cases}$$

for $a, b > 0$ and $c \in R$, Davidovich (1992) established some bounds for the difference between the two cdf's. Namely,

$$F_X\left(x; b + c, \frac{b}{a}\right) - G_X(x;a,b,c) < \begin{cases} e^{-a} & \text{if } x \le c, \\ \dfrac{2e^{-2}}{a-2} & \text{if } c \le x \le c + 2b, \\ e^{a-2^a} & \text{if } x \ge c + 2b. \end{cases}$$

If $Y_1, Y_2, \ldots,$ are independent variables, each having the exponential distribution

$$\Pr[Y \le y] = 1 - e^{-y}, \qquad y > 0, \tag{1.18}$$

and if L is the zero-truncated Poisson variable

$$\Pr[L = l] = \frac{(e^\lambda - 1)^{-1}\lambda^l}{l!}, \qquad l = 1, 2, \ldots, \tag{1.19}$$

then the random variable defined by

$$X = \max(Y_1, \ldots, Y_L)$$

has the extreme value distribution with cdf

$$\Pr(X \le x) = (e^\lambda - 1)^{-1}[\exp\{\lambda(1 - e^{-x})\}] = c\exp[-\lambda e^{-x}]. \tag{1.20}$$

In connection with the asymptotic nature of extreme value distributions (and densities), the following recent result by Beirlant and Devroye (1999) may be of relevance.

Let $X_1, \ldots, X_n$ be an i.i.d. sample drawn from a density f (with cdf F) on the real line. The basic tenet of the classical extreme-value distributional theory is that $Y_n = \max(X_1, \ldots, X_n)$ is in the domain of attraction of an extreme value distribution and a few results on the rate of convergence of the distribution of Y_n to its limit distribution in uniform metric and total variation distance are available (de Haan and Resnick (1996)).

Various methods have been developed to test whether a sequence of i.i.d. observations belongs to the domain of attraction of one of the three distributions (see Castillo (1986), Marohn (1998) and Alves and Gomes (1996) for the Gumbel type; Tiku and Singh (1981) and Shapiro and Brain (1987) for the Weibull type).

However, the general problem of designing an estimate f_{Y_n} of the density of Y_n (respectively the cdf of Y_n) that is consistent in total variation is unsolvable. Specifically, Beirlant and Devroye (1999) showed that there exists a unimodal infinitely many times differential density such that

$$\inf_n E\{|f_n(x) - g_n(x)|\}dx \geq \frac{1}{49};$$

here $g_n = nfF^{n-1}$ is the density of Y_n. Thus a universally consistent density estimator does not exist. Hence to study rates of convergences of Y_n to its limit distribution, both tail and smoothness conditions are required.

Angus (1993) considered asymptotic analysis of bootstrap distributions for the extremes from an i.i.d. sample. In contrast to the case of almost sure convergence to a fixed (normal) distribution in the case of the sample mean (a finite variance case), the bootstrap distribution of an extreme tends in distribution to a *random* probability measure.

1.4 Distribution Function and Moments of Type 1 Distribution

Corresponding to type 1 distribution (1.1) is the probability density function:

$$p_X(x) = \sigma^{-1}e^{-(x-\mu)/\sigma}\exp\left[-e^{-(x-\mu)/\sigma}\right]. \tag{1.21}$$

We reiterate that

$$-\log\{-\log F_X(x;\mu,\sigma)\} = (x-\mu)/\sigma$$

and the cdf would be linear if drawn on graph paper on which the percentage scale were doubly logarithmic. The empirical cdf would then be approximately linear. Specially prepared probability graph paper for this purpose is commercially available. We shall discuss applications of this graph paper in the section on statistical inference.

If $\mu = 0$ and $\sigma = 1$, or equivalently, the distribution is of $Y = (X-\mu)/\sigma$, we have the *standard form*

$$p_Y(y) = \exp\left[-y - e^{-y}\right]. \tag{1.22}$$

Since, the variable $Z = \exp[-(X-\mu)/\sigma] = e^{-Y}$ has the exponential distribution

$$p_Z(z) = e^{-z}, \qquad z \geq 0,$$

it follows that

$$E[e^{t(X-\mu)/\sigma}] = E(Z^{-t}) = \Gamma(1-t)$$

for $t < 1$. The moment generating function of X is

$$E[e^{tX}] = e^{t\mu}\Gamma(1-\sigma t), \qquad \sigma|t| < 1, \tag{1.23}$$

and the cumulant generating function is

$$\Psi(t) = \mu t - \log\Gamma(1-\sigma t). \tag{1.24}$$

Thus, the first order cumulant

$$k_1(X) = E(X) = \mu - \sigma\psi(1) = \mu + \gamma\sigma$$
$$= \mu + 0.57722\sigma\,, \tag{1.25}$$

where γ is Euler's constant, and the higher cumulants are

$$k_i(X) = (-\sigma)^i \psi^{(i-1)}(1)\,, \qquad i \geq 2\,, \tag{1.26}$$

where $\psi(\cdot)$ is the digamma function.

The variance is:

$$\text{Var}(X) = \frac{1}{6}\pi^2\sigma^2 = 1.64493\sigma^2 \tag{1.27a}$$

and

$$\text{std. dev.}(X) = 1.28255\sigma\,. \tag{1.27b}$$

The moment ratios are

$$\alpha_3^2(X) \equiv \beta_1(X) \simeq 1.29857\,,$$
$$\alpha_4(X) = \beta_2(X) = 5.4\,. \tag{1.28}$$

We emphasize that μ and σ are (purely) location and scale parameters, respectively.

All distributions (1.21) have the same shape.

The distribution is unimodal. Its mode is at $X = \mu$ and there are points of inflection at

$$X = \mu \pm \sigma \log\left[\frac{1}{2}(3 + \sqrt{5})\right] \simeq \mu \pm 0.96242\sigma\,. \tag{1.29}$$

From (1.1) for $0 < p < 1$, the pth quantile defined by $F(X_p) = p$ becomes

$$X_p = \mu - \sigma \log(-\log p)\,. \tag{1.30}$$

From (1.30) we immediately obtain the lower quartile, median, and upper quartile to be

$$X_{0.25} = \mu - 0.32663\sigma\,, \tag{1.31}$$

$$X_{0.50} = \mu + 0.36611\sigma\,, \tag{1.32}$$

$$X_{0.75} = \mu + 1.24590\sigma\,, \tag{1.33}$$

respectively.

Quantiles of the distribution are easy to compute from (1.30). They are of special importance in applications of extreme value distributions. Most of the standard distribution (1.22) is contained in the interval $(-2, 7)$. For the type 1 distribution function we find the probability between $\mu - 2\sigma$ and $\mu + 7\sigma$ to be 0.998. That is 99.8% of the distribution lies between Mean $-2.0094 \times$ (standard deviation) and Mean $+$ $5.0078 \times$ (standard deviation).

Table 1.1 gives *standardized* percentile points (i.e. for a type 1 extreme value distribution with expected value zero and standard deviation 1, corresponding to $\sigma = \sqrt{6}/\pi = 0.77970$; $\mu = -\gamma\sigma = -0.45006$).

The positive skewness of the distribution is clearly indicated by these values. See Fig. 1.1. We must emphasize that often type 1 is chosen without further investigation for practical (ease of fitting) as well as theoretical reasons. Cohen (1982) shows that perhaps this preference for type 1 may sometimes be misguided (see the Appendix).

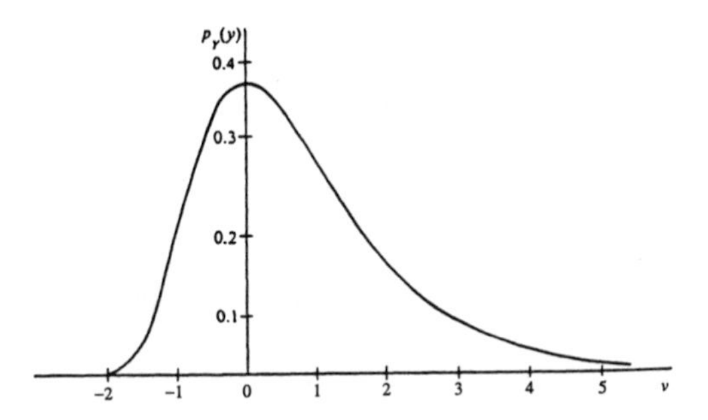

Figure 1.1: Standard type 1 probability density function (1.21).

Table 1.1: Standardized percentiles for type 1 extreme value distribution.

α	Percentiles
0.0005	−2.0325
0.0001	−1.9569
0.005	−1.7501
0.01	−1.6408
0.05	−1.3055
0.1	−1.1004
0.9	1.3046
0.95	1.8658
0.99	3.1367
0.9975	4.2205
0.999	4.9355

1.5 Order Statistics, Record Values and Characterizations

If $Y_1' \leq Y_2' \leq Y_n'$ are the order statistics corresponding to n independent random variables each having the standard type 1 extreme value distribution, then the probability density function of $Y_r'(1 \leq r \leq n)$ is

$$p_{Y_r'}(y) = \frac{n!}{(r-1)!(n-r)!} \sum_{j=0}^{n-r} (-1)^j \binom{n-r}{j} e^{-y-(j+r)e^{-y}}, \qquad -\infty < y < \infty. \quad (1.34)$$

From (1.34) the kth moment of Y_r' can be written as

$$E[Y_r'^k] = \frac{n!}{(r-1)!(n-r)!} \sum_{j=0}^{n-r} (-1)^j \binom{n-r}{j} g_k(r+j), \quad (1.35)$$

where

$$g_k(c) = \int_{-\infty}^{\infty} y^k e^{-y-ce^{-y}} dy$$
$$= (-1)^k \int_{-\infty}^{\infty} (\log u)^k e^{-cu}\, du \qquad (\text{with } u = e^{-y})\,. \tag{1.36}$$

The functions $g_1(\cdot)$ and $g_2(\cdot)$ required for the expressions of the first two moments of order statistics are

$$g_1(c) = -\frac{\Gamma'(1)}{c} + \frac{\Gamma(1)}{c}\log c = \frac{1}{c}(\gamma + \log c) \tag{1.37}$$

and

$$g_2(c) = \frac{1}{c}\left\{\frac{\pi^2}{6} + (\gamma + \log c)^2\right\}; \tag{1.38}$$

as above (here γ is Euler's constant).

Proceeding similarly the product moment of Y'_r and Y'_s $(1 \le r < s \le n)$ can be shown to be

$$E[Y'_rY'_s] = \frac{n!}{(r-1)!(s-r-1)!(n-s)!}$$
$$\times \sum_{i=0}^{s-r-1}\sum_{j=0}^{n-s}(-1)^{i+j}\binom{s-r-1}{i}\binom{n-s}{j}$$
$$\times\, \phi(r+i, s-r-i+j)\,, \tag{1.39}$$

where the function ϕ is the double integral

$$\phi(t,u) = \int_{-\infty}^{\infty}\int_{-\infty}^{y} xye^{x-te^x}e^{y-ue^y}\,dx\,dy\,, \qquad t,u>0\,. \tag{1.40}$$

Lieblein (1953) derived an explicit expression for the ϕ function in (1.40) in terms of Spence's function which has been tabulated by Abramowitz and Stegun (1965) and other handbooks.

Balakrishnan and Chan (1992a) presented tables of means, variances and covariances of all order statistics for sample sizes $n = 1(1)15(5)30$. Complete tables for all sample sizes up to 30 have also been prepared by Balakrishnan and Chan (1992c). Mahmoud and Ragab (1975) and Provasi (1987) have provided further discussions on order statistics.

Suppose that $Y_1, Y_2, \ldots$ is a sequence of i.i.d. standard type 1 extreme value random variables and that $Y_{L(1)} \equiv Y_1$, $Y_{L(2)}, \ldots$ are the corresponding lower record values. That is, $L(1) \equiv 1$ and $L(n) = \min\{j : j > L(n-1),\ Y_j < Y_{L(n-1)}\}$ for $n = 2, 3, \ldots$, $\{Y_{L(n)}\}_{j=1}^{\infty}$ form the lower record value sequence. Then the density function of $Y_{L(n)}, n \ge 1$, is given by

$$p_{Y_{L(n)}}(y) = \frac{1}{(n-1)!}\{-\log F_Y(y)\}^{n-1}p_Y(y)$$
$$= \frac{1}{(n-1)!}e^{-ny}e^{-e^{-y}}, \qquad -\infty < y < \infty\,. \tag{1.41}$$

This is the density function of the so-called log-gamma population when the shape parameter $\kappa = n$. It will be discussed below in the section on Related Distributions.

Thus, for $n = 1, 2, \ldots,$

$$E[Y_{L(n)}] = \gamma - \sum_{i=1}^{n-1} \frac{1}{i}, \qquad \text{var}(Y_{L(n)}) = \frac{\pi^2}{6} - \sum_{i=1}^{n-1} \frac{1}{i^2}, \tag{1.42}$$

and

$$E(Y_{L(n+1)}) = E(Y_{L(n)}) - \frac{1}{n},$$

$$E(Y_{L(n+1)}) = E(Y_{L(m)}) - \sum_{p=m}^{n} \frac{1}{p} \qquad \text{(see below)}. \tag{1.42a}$$

The joint density function of $Y_{L(m)}$ and $Y_{L(n)}$, $1 \le m \le n$, is given by

$$\begin{aligned} p_{Y_{L(m)},Y_{L(n)}}(y_1, y_2) &= \frac{1}{(m-1)!(n-m-1)!}\{-\log F_Y(y_1)\}^{m-1} \frac{p_Y(y_1)}{F_Y(y_1)} \\ &\quad \times \{-\log F_Y(y_2) + \log F_Y(y_1)\}^{n-m-1} p_Y(y_2) \\ &= \frac{1}{(m-1)!(n-m-1)!} e^{-my_1}(e^{-y_2} - e^{-y_1})^{n-m-1} e^{-y_2} e^{-e^{-y_2}}, \\ &\qquad -\infty < y_2 < y_1 < \infty. \end{aligned} \tag{1.43}$$

Upon writing the joint density of $Y_{L(m)}$ and $Y_{L(n)}$, $1 \le m \le n$, in (1.43) as

$$\begin{aligned} P_{Y_{L(m)},Y_{L(n)}}(y_1, y_2) &= \frac{(n-1)!}{(m-1)!(n-m-1)!} e^{-m(y_1-y_2)}(1 - e^{-(\nu_1-\nu_2)})^{n-m-1} \\ &\quad \times \frac{1}{(n-1)!} e^{-ny_2} e^{-e^{-v_2}}, \quad -\infty < y_2 < y_1 < \infty, \end{aligned} \tag{1.44}$$

we observe that $Y_{L(m)} - Y_{L(n)}$ and $Y_{L(n)}$ for $(1 \le m \le n)$ are statistically independent. As a result, we get

$$\text{cov}(Y_{L(m)}, Y_{L(n)}) = \text{var}(Y_{L(n)}) = \frac{\pi^2}{6} - \sum_{i-1}^{n-1} \frac{1}{i^2}. \tag{1.45}$$

These properties are similar to those of order statistics arising from standard exponential random variables. In fact, it follows from (1.44) that $Y_{L(m)} - Y_{L(n)}$ is distributed as the $(n-m)$th-order statistic in a sample of size $n-1$ from the standard exponential distribution, say $Z_{n-m:n-1}$. For the special case when $m = 1$, we then have $Y_{L(1)} - Y_{L(n)} = Y_1 - Y_{L(n)} = Z_{n-1:n-1}$. Suppose that $X'_{i:j}$ is the ith-order statistic in a random sample of size j from a distribution $F(\cdot)$. If the distribution function of $(X'_{j:j} - a_j)/b_j$ converges weakly to a nondegenerate distribution function $G(\cdot)$ as $j \to \infty$ for sequences of constants a_j and positive b_j, then Nagaraja (1982) showed that the joint distribution function of $(X'_{j-i+1:j} - a_j)/b_j$ $1 \le i \le n$, converges to that of $X_{L(i)}$, $1 \le i \le n$.

Recurrence relations for the single and product moments of the lower record values were obtained by Ahsanullah (1994). A useful result is the following:

$$\text{For } n \ge 1 \quad \text{and} \quad r = 0, 1, 2, \ldots,$$

$$E(X_{L(n+1)}^{r+1}) = E(X_{L(n)}^{r+1}) - \frac{r+1}{n} E(X_{L(n)}^{r}) \qquad \text{(compare with (1.42a))}.$$

As mentioned earlier, X has a type 1 extreme value distribution if and only if e^X has a Weibull distribution, and $e^{X/\sigma}$ has an exponential distribution, and $\exp\{(X-\mu)/\sigma\}$ has a *standard* exponential distribution.

Some characterization theorems for exponential distributions may also be used for type 1 extreme value distributions, simply by applying them to $e^{X/\sigma}$, or $\exp\{(X-\mu)/\sigma\}$. Dubey (1966) characterized this distribution by the property that $Y_n = \min(X_1, X_2, \ldots, X_n)$ is a type 1 random variable if and only if $X_1, X_2, \ldots, X_n$ are independent identically distributed type 1 random variables.

Sethuraman (1965) has obtained revealing structural characterizations of all three types of extreme value distributions, in terms of "complete confounding" of random variables. If X and Y are independent and the distributions of Z, given $Z = X$, and Z given $Z = Y$ are the same [e.g., Z might be equal to $\min(X, Y)$ as in the case described in Sethuraman (1965)], they are said to *completely confound* each other with respect to the third. Sethuraman showed that if all pairs from the variables X, Y and Z completely confound each other with respect to the third and if Y, Z have the same distributions as $a_1X + b_1$, $a_2X + b_2$, respectively [with $(a_1, b_1) \neq (a_2, b_2)$], then the distribution of X is one of the three extreme value (minimum) distributions (provided we limit ourselves to the cases when $\Pr[X > Y] > 0$; $\Pr[Y > X] > 0$, etc.). The type of distribution depends on the values of a_1, a_2, b_1, b_2.

There are a number of characterizations of the type 1 distribution in the framework of extreme value theory. The most prominent one is that the type 1 distribution is the only *max-stable* probability distribution function with the entire real line as its support; see e.g. Theorem 1.4.1 in Leadbetter, Lindgren, and Rootzén (1983). The concept of max-stability is of special usefulness especially for the multivariate extreme value distributions; see Chap. 3. In addition to the characterizations of the type 1 distribution itself, there are several characterization results available for the maximal domain of attraction of the type 1 distribution; de Haan (1970) is a good initial source of information on this as well as on characterizations for type 2 and type 3 distributions.

Tikhov (1991) has characterized the extreme value distributions by the limiting information quantity associated with the maximum likelihood estimator based on a multiply censored sample.

1.6 Generation, Tables, Probability Paper, Plots and Goodness of Fit

Collection of tables cited below are of more than just historical interest.

The following tables are included in Gumbel (1953):

(a) Values of the standard cumulative distribution function, $F = \exp(-e^{-y})$, and probability density function, $\exp(-y - e^{-y})$, to seven decimal places for $y = -3(0.1) - 2.4(0.05)0.00(0.1)4.0(0.2)8.0(0.5)17.0$.

(b) The inverse of the cumulative distribution function (percentiles), $y = -\log(-\log F)$ to five decimal places for

$$F = 0.0001(0.0001)0.0050(0.0001)0.988(0.0001)0.9994(0.00001)0.99999.$$

In Owen's tables (1962) there is a similar table, to four decimal places for

$$F = 0.0001(0.0001)0.0010(0.0010)0.0100(0.005)0.100(0.010)0.90(0.005)$$
$$0.990(0.001)0.999(0.0001)1 - 10^{4(1)7}, 1 - 1/2 \cdot 10^{-4(1)7}.$$

The special interest in very high value of F, by both Gumbel (1953) and Owen (1962), is associated with the genesis of the distribution.

From (1.1) it follows that

$$-\log(-\log \Pr[X < x]) = (x - \mu)/\sigma. \tag{1.46}$$

As it has already been pointed out, but repeated here due to practical importance, if the cumulative observed relative frequency F_x — equal to (number of observations less than or equal to x)/(total number of observations) — is calculated, and $-\log(-\log F_x)$ plotted against x, an approximately straight line relation should be obtained, with slope σ^{-1} and intersecting the horizontal (x) axis at $x = \mu$. In using graph paper with a vertical scale that gives $-\log(-\log F_x)$ directly, it is not necessary to refer to tables of logarithms. Such graph paper is sometimes called *extreme value probability paper* or *Gumbel paper*. Equally it is also common to use such paper with the x-axis vertical, and for practical purposes it is sometimes convenient to have the $-\log(-\log F)$ marked not with F_x but with the "return period" $(1 - F_x)^{-1}$; see e.g. Gumbel (1949a) and Kimball (1960). Such a paper is called *extreme probability paper*.

Evidently, pseudorandom numbers from the standard type 1 distribution may be generated easily either through the inverse cdf method along with an efficient uniform random generator or through the relationship with the exponential distribution along with an efficient exponential random generator. Sibuya (1967) has discussed the latter. Landwehr, Matalas, and Wallis (1979) have advocated the use of the Lewis–Goodman–Miller algorithm for generating pseudorandom numbers from the uniform distribution for this purpose. In a similar manner, the Fréchet distribution can be generated from the Pareto distribution and the Weibull from the power function distribution.

Due to the prominence and significance of the extreme value distributions, considerable work has been done with regard to testing whether an extreme value distribution is appropriate for the data at hand. The book by D'Agostino and Stephens (1986) provides an elaborate account of various goodness-of-fit tests developed for the extreme value distributions.

Consider type 1 extreme value distribution:

$$\Pr[Y \geq y] = F(y; \mu, \sigma) = 1 - \exp\left[-\exp\left(\frac{y-\mu}{\sigma}\right)\right], \qquad -\infty \leq y \leq \infty,$$

where (μ, σ) = (location, scale) parameter.

As already mentioned on several occasions, taking the logarithm we obtain the linear relationship

$$y = \mu + \sigma \ln[-\ln(1 - F(y; \mu, \sigma))]$$

for plotting the observed data on type 1 extreme value (or Gumbel) paper.

If y_i, $i = 1, \ldots, n$, is a set of *ordered* observations and p_i, $i = 1, \ldots, n$, is a set of plotting positions given by a plotting method, then the plotted points, $(y_i, \ln[-\ln(1 - p_i)]$,

$i = 1, \dots, n$) should be approximately distributed on a straight line on a type 1 extreme value probability paper (provided the data reasonably fit a type 1 extreme value distribution).

Shimokawa and Liao (1999) examined three plotting methods, the so-called graphical plotting techniques (GPT):

(a) Median ranks. Here p_i is defined by

$$\frac{1}{2} = i \cdot \binom{n}{i} \int_0^{p_i} t^{i-1}(1-t)^{n-i}\,dt\,.$$

(b) Mean ranks:

$$p_i = \frac{i}{n+1}\,.$$

(c) Symmetrical sample cdf ranks:

$$p_i = \frac{i-0.5}{n}\,.$$

The authors combine the least square method (LSM) and the plotting procedure to estimate μ and σ. Specifically let $c_i = \ln\left[-\ln\left(1-p_i\right)\right]$ and μ and σ be obtained by minimizing

$$T = \sum_{i=1}^{n} (y_i - \hat{\mu} - \hat{\sigma} c_i)^2 \qquad \text{for given } y_i \text{ and } c_i\,,$$

yielding

$$\hat{\sigma} = \left[\sum_{i=1}^{n} (y_i - \bar{y}) \cdot (c_i - \bar{c})\right] \Big/ \left[\sum_{i=1}^{n} (c_i - \bar{c})^2\right],$$

$$\hat{\mu} = \bar{y} - \sigma \cdot \bar{c}\,,$$

where

$$\bar{c} = \left(\frac{1}{n}\right) \cdot \sum_{i=1}^{n} c_i\,, \qquad \bar{y} = \left(\frac{1}{n}\right) \cdot \sum_{i=1}^{n} y_i\,.$$

Monte–Carlo simulation was used to obtain critical values for the Kolmogorov–Smirnov statistics (K-S), Cramér–von Mises statistics (C-M) and the Anderson–Darling (A-D) statistics for goodness-of-fit tests when population parameters are estimated from complete samples by graphical plotting techniques.

10^6 sets of samples for each sample size of 3(1)20, 25(5)50 and 60(10)100 were generated. The power was investigated for the 3 GPTs and for maximum likelihood estimator (MLE). The simulation provided power results using 10^4 repetitions for each sample size of 5, 10, 25 and 40.

Let:

(a) $x_1 < x_2 < \cdots < x_n$ be order statistics for a sample of size n from a population defined by a continuous cdf $F(x)$.
(b) $F_0(x;\theta)$ be a specified family of models that contain a set of parameters θ.

A goodness-of-fit test is used to test the null hypothesis

$$H_0 : F(x) = F_0(x; \theta) .$$

The K-S statistic for a goodness-of-fit test is based on

$$D_n = \max_{1 \le i \le n} [\delta_i] ,$$

where

$$\delta_i \equiv \left[\frac{i}{n} - F_0(x_i; \theta), F_0(x_i; \theta) - \frac{i-1}{n} \right] .$$

The C-M statistic is represented by:

$$W_n^2 = \sum_{i=1}^{n} \left[F_0(x_i; \theta) - \frac{i - 0.5}{n} \right]^2 + \frac{1}{12n} .$$

A computational formula of the A-D statistic is:

$$A_n^2 = - \sum_{i=1}^{n} \frac{2i-1}{n} \cdot [\ln (F_0(x_i; \theta)) + \ln (1 - F_0(x_{n+1-i}; \theta))] - n .$$

Since, in practice, $F_0(x; \theta)$ often depends on unknown parameters, it must be modified to $F_0(x; \hat{\theta})$ by using parameter estimates. Then, the K-S, C-M and A-D statistics are modified to $\hat{D}_n$, $\hat{W}_n^2$, $\hat{A}_n^2$, respectively.

Conclusions of the study by Shimokawa and Liao (1999) as applied to type 1 distribution are:

(a) Among the three GPTs, the symmetrical ranks give more powerful results than the median and mean ranks for the K-S, C-M and A-D statistics. Symmetrical ranks provide more powerful results than the MLE for the K-S and A-D statistics. For the C-M statistic, the MLE provides more powerful results than the three GPTs.

(b) Generally, the A-D statistic coupled with the symmetrical ranks and the least square method of estimation (LSM) is most powerful among the competitors in this study and is recommended for practical use.

One of the easiest goodness-of-fit tests is the "correlation coefficient" test for the type 1 extreme value distribution. This test is based on the product-moment correlation between the sample order statistics and their expected values. Since $E[X_i'] = \mu + \sigma E[Y_i']$, one may as well use the correlation between the sample order statistics X_i' and the expected values of *standard* order statistics $E[Y_i']$, for the type 1 extreme value distribution. Naturally large values (close to 1) of this correlation will support the assumption of the type 1 extreme value distribution for the data at hand. Smith and Bain (1976) discussed this test and presented tables of critical points; tables were also provided by these authors for the case when the available sample is Type-II censored. A more extensive table of points for $n(1 - R^2)$, where – R is the sample correlation coefficient, has been constructed by Stephens (1986). Kinnison (1989) discussed the same correlation test for the type 1 extreme value distribution and presented tables of smoothed values of the percentage points of R (in the case of complete samples) for $n = 5(5)30(10)100$, 200.

As aptly mentioned by Lockhart and Spinelli (1990) even though the correlation test is simple to use and has an intuitive appeal, its power properties are undesirable. Indeed, McLaren and Lockhart (1987) have shown that the correlation test has asymptotic efficiency equal to 0 relative to K-S, C-M and A-D tests discussed above.

Earlier, Stephens (1977) presented goodness-of-fit tests based on empirical distribution function statistics W^2, U^2 and A^2 given by

$$W^2 = \sum_i \left\{ F_X(X_i') - \frac{2i-1}{2n} \right\}^2 + \frac{1}{12n} \tag{1.47}$$

(cf. above),

$$U^2 = W^2 - n\left\{ \frac{1}{n} \sum_i F_X(X_i') - \frac{1}{2} \right\}^2 \tag{1.48}$$

and

$$A^2 = -\frac{1}{n} \sum_i (2i-1)[\log F_X(X_i') + \log\{1 - F_X(X_{n-i+1}')\}] - n \tag{1.49}$$

(cf. above).

This author discusses the asymptotic percentage points of these three statistics for the three cases when one or both of the parameters μ and σ need to be estimated from the data (using the MLEs). He also suggested slight modifications of these statistics in order to enable the usage of the asymptotic percentage points in case of small sample sizes.

Along similar lines, Chandra, Singpurwalla and Stephens (1981) considered the K-S statistics D^+, D^- and D and the Kuiper statistic V given by:

$$D^+ = \max_i \left\{ \frac{i}{n} - F_X(X_i') \right\}. \tag{1.50}$$

$$D^- = \max_i \left\{ F_X(X_i') - \frac{i-1}{n} \right\}, \tag{1.51}$$

$$D = \max(D^+, D^-) \tag{1.52}$$

and

$$V = D^+ + D^- . \tag{1.53}$$

They determined some percentage points of these statistics for the three cases when one or both of the parameters μ and σ need to be estimated from the data (using the MLEs). A stabilized probability plot proposed by Michael (1983) is to plot

$$S_i = \frac{2}{\pi} \sin^{-1} \left\{ F_X\left(\frac{X_i' - \mu}{\sigma} \right) \right\}^{1/2} \tag{1.54}$$

with respect to

$$r_i = \frac{2}{\pi} \sin^{-1} \left(\frac{i - 0.5}{n} \right)^{1/2} .$$

In this way the unequal variance problem of the plotted points can be avoided, since S_i in (1.54) have approximately equal variance, as the asymptotic variance of $\sqrt{n}S_i$ is the constant $(1/\pi^2)$. A goodness-of-fit statistic that arises naturally from the stabilized probability plot is

$$D_{\text{sp}} = \max_i |r_i - S_i| . \tag{1.55}$$

Kimber (1985) presented critical values for the statistic $D_{\rm sp}$ in (1.55) for selected choices of n. Van Montfort (1973) dealt with testing goodness-of-fit for the type 1 distribution of largest extremes, where $\Pr(X < x) = \exp(-x-\mu)/\sigma$ with unknown location parameter μ $(-\infty < \mu < +\infty)$ and scale parameter σ $(0 < \sigma < \infty)$. The proposed test is intended to have a high power against the alternative distribution, $\Pr(\log(X-\nu) < x) = \exp(-\exp(-(x-\mu)/\sigma))$, where ν is the lower bound of the support of X. The statistic is a function of standardized spacings. Critical values and power are approximated by means of Monte–Carlo methods.

Tsujitani, Ohta and Kase (1980) proposed a test based on the sample entropy, presented its critical points for some sample sizes determined through Monte–Carlo simulations, and showed that it has desirable power properties compared with some of the tests mentioned above. Öztürk (1986) considered the Shapiro–Wilk W test and presented some percentage points determined through Monte–Carlo simulations (see below). A modification of the W statistic has been considered by Öztürk and Korukoğlu (1988) in which the test statistic has been obtained as the ratio of two linear estimators of the parameter. These authors have determined percentage points of this statistic through Monte–Carlo simulations and have also shown by means of an empirical comparative study that this test possesses good power properties.

With the normalized spacings

$$Z_i = \frac{X'_{i+1} - X'_i}{E(Y'_{i+1}) - E(Y'_i)}, \qquad i = r+1, \ldots, n-s-1,$$

where Y'_i are order statistics from the standard distribution, and

$$Z^*_i = \frac{\sum_{j=r+1}^{i} Z_j}{\sum_{j=r+1}^{i+1} Z_j}, \qquad i = r+1, \ldots, n-s-2, \tag{1.56}$$

Lockhart *et al.* (1986b) focus on the A-D statistic

$$\begin{aligned} A^2 &= -(n-r-s-2) \\ &\quad - \frac{1}{n-r-s-2}\left[\sum_{i=r+1}^{n-s-2} (2i-1)\{\log Z^*_i + \log(1 - Z^*_{n-s-1-i})\}\right] \end{aligned} \tag{1.57}$$

discussed above and compare its performance with the S-statistic introduced by Mann, Scheuer and Fertig (1973) and the $\bar{Z}^*$ statistic introduced by Tiku and Singh (1981). Here, $T = 1 - Z^*_t$, where

$$t = \begin{cases} r + \dfrac{n-r-s}{2} & \text{if } n-r-s \text{ is even} \\ r + \dfrac{n-r-s-1}{2} & \text{if } n-r-s \text{ is odd}\ , \end{cases} \tag{1.58}$$

and

$$\bar{Z}^* = \frac{1}{n-r-s-2}\sum_{i=r+1}^{n-s-2} Z^*_i\,. \tag{1.59}$$

In agreement with the more recent conclusions of Shimokawa and Liao (1999), Lockhart *et al.* (1986b), based on their comparative study, recommend overall the A^2 test, and they also mentioned that while the $\bar{Z}^*$ test gives good power in many situations, it may occasionally be inconsistent.

Hasofer and Wang (1992) proposed a statistic to test the hypothesis that a sample comes from a distribution in the domain of attraction of the type 1 distribution. It is based on the top k order statistics and is a generalization of the well-known Shapiro–Wilk goodness-of-fit statistic (W). The critical region of the test and its power against the alternative that the sample comes from a distribution in another domain of attraction are studied theoretically and by simulation.

Suppose that $F(x)$ is in the domain of attraction of the Gumbel distribution; i.e. there are two sequences $\{a_n\}$ and $\{b_n\}$ such that

$$\lim_{n\to\infty} F_n(a_n x + b_n) = \exp\{-\exp\{-x\}\}, \qquad -\infty < x < \infty .$$

Let $X_{1n} \geq \cdots \geq X_{kn} \geq \cdots \geq X_{nn}$ be the order statistics of a sample of size n from $F(x)$ and $Z_{in} = (X_{in} - b_n)/a_n$. Consider now the random vector $Z_{kn}^T = (Z_{1n}, \ldots, Z_{kn})$, where k is a fixed number. It has been shown that as $n \to \infty$, the random vector Z_{kn}^T converges in distribution to the limit random vector $U_k^T = (U_1, \ldots, U_k)$, where the U_i's have the joint density function $h(u_1, \ldots, u_k) = \exp\{-\exp\{-u_k\} - \sum_{i=1}^k u_i\}$, $u_1 \geq \cdots \geq u_k$. Hasofer and Wang (1992) showed that under the null hypothesis that the joint distribution of Z_k is $h(u_1, \ldots, u_k)$, the above-mentioned classical goodness-of-fit criterion

$$W = \frac{k(\bar{X} - X_k)}{(k-1)\left[\sum_{i=1}^{k}(X_i - \bar{X})^2\right]}$$

proposed by Shapiro and Wilk (1965) for testing normality is, in this case, a simple function of the so-called Greenwood statistic G_2 introduced over 50 years ago (Greenwood (1946)) based on differences of order statistics. (See also Moran (1947), (1953).) The authors have shown that the distribution of W for this model shifts to the left for the Fréchet (Type 2) distributions and towards the right for the Weibull (Type 3) distributions.

Utilizing Kimball's (1956) simplified linear estimators $\hat{\mu}$ and $\hat{\sigma}$ to be discussed in the next section, Aly and Shayib (1992) proposed the statistic

$$M_n = -\sum_{i=1}^{n}\left\{\left(\frac{X_i' - \hat{\mu}}{\hat{\sigma}}\right) - \log\left[-\log\left(1 - \frac{i}{n+1}\right)\right]\right\}^2 \times \left(1 - \frac{i}{n+1}\right)\log\left(1 - \frac{i}{n+1}\right) \qquad (1.60)$$

for testing the validity of the type 1 extreme value distribution for the minimum. They determined the critical points of M_n for selected sample sizes through Monte–Carlo simulations. Aly and Shayib (1992) also compared the power of this test with some other tests including the A^2 test in (1.57). From their brief power study it seems that the M_n test outperforms the A^2 test for skewed alternatives; however, in the case of symmetric alternatives, the A^2 test seems to be considerably better than the M_n test.

Tiago de Oliveira (1981) discussed the *statistical* choice among the different extreme value models. Vogel (1986) discussed further the probability plot and the associated correlation coefficient test. Cohen (1986, 1988) presented detailed critical discussions on the large-sample theory for fitting extreme value distributions to maxima.

1.7 Methods of Inference

This section is rather lengthy. We follow E. J. Gumbel's (1958) dictum that "no distribution should be stated without an explanation of how the parameters are estimated even at the risk that the methods used will not stand up to the present rigorous requirements of mathematically minded statisticians".

Let $X_1, X_2, \ldots, X_n$ be a random sample of size n from the type 1 extreme value distribution in (1.21). Then as Downton (1966) has shown, the Cramér–Rao lower bounds of variances of unbiased estimators of μ and σ are given by

$$\begin{aligned} \{1+6(1-\gamma)^2\pi^{-2}\}\sigma^2 n^{-1} &= 1.10867\sigma^2 n^{-1}, \\ 6\pi^{-2}\sigma^2 n^{-1} &= 0.60793\sigma^2 n^{-1}, \end{aligned} \tag{1.61}$$

respectively.

As has already been mentioned, if Z has a Weibull distribution with probability density function

$$p_Z(z) = \frac{c}{\sigma}\left(\frac{z-\mu}{\sigma}\right)^{c-1} e^{-[(z-\mu)/\sigma]^c}, \qquad z \geq \mu, \tag{1.62}$$

then $\log(Z-\mu)$ has a type 1 extreme value distribution. Consequently, if μ is known, the methods of estimation discussed in this section for the type 1 extreme value distribution can also be used for estimating the parameters σ and c of the Weibull distribution (1.62) and vice versa.

1.7.1 *Moment Estimation*

This is one of the most popular methods of estimating parameters. Let $\bar{X}$ and S denote the sample mean and the sample standard deviation. Then using Eqs. (1.25) and (1.27), we simply obtain the moment estimates of μ and σ as

$$\tilde{\mu} = \frac{\sqrt{6}}{\pi}S \quad \text{and} \quad \begin{aligned} \tilde{\sigma} &= \bar{X} - \gamma\tilde{\mu} \\ &= \bar{X} - 0.450041S \end{aligned} \tag{1.63}$$

(see Lowery and Nash (1970), Landwehr *et al.* (1979)).

Tiago de Oliveira (1963) has shown that

$$\mathrm{var}(\tilde{\mu}) \simeq \frac{\sigma^2}{n}\left\{\frac{\pi^2}{6} + \frac{\gamma^2}{4}(\beta_2 - 1) - \frac{\pi}{\sqrt{6}}\gamma\sqrt{\beta_1}\right\} \tag{1.64}$$

and that

$$\mathrm{var}(\tilde{\sigma}) = \frac{\sigma^2}{4n}(\beta_2 - 1), \tag{1.65}$$

where β_1 and β_2 are the coefficients of skewness and kurtosis as given in (1.28). Substituting for their values, we get

$$\mathrm{var}(\tilde{\mu}) \simeq \frac{1.1678\sigma^2}{n} \quad \text{and} \quad \mathrm{var}(\tilde{\sigma}) \simeq \frac{1.1\sigma^2}{n}. \tag{1.66}$$

A comparison of the variance formulas in (1.66) with the Cramér–Rao lower bounds in (1.61) readily reveals that the moment estimator $\tilde{\mu}$ has about 95% efficiency while the moment estimator $\tilde{\sigma}$ has only about 55% efficiency. The estimators $\tilde{\mu}$ and $\tilde{\sigma}$ are both $\sqrt{n}$-consistent; i.e. $\sqrt{n}\,(\tilde{\mu}-\mu)$ and $\sqrt{n}\,(\tilde{\sigma}-\sigma)$ are bounded in probability.

Tiago de Oliveira (1963) has shown that the joint asymptotic distribution of $\tilde{\mu}$ and $\tilde{\sigma}$ is bivariate normal with mean vector $(\mu,\sigma)'$, variances as given in (1.65) and the correlation coefficient given by

$$\rho_{\tilde{\mu},\tilde{\sigma}} = \frac{\pi^2|\sqrt{\beta_1} - 3\gamma(\beta_2 - 1)/2\pi]/6}{[\{\pi^2/6 + \gamma^2(\beta_2 - 1)/4 - \pi(\gamma\sqrt{\beta_1})/\sqrt{6}\}(\beta_2 - 1)]^{1/2}} \simeq 0.123\,. \tag{1.67}$$

Using this asymptotic result, asymptotic confidence regions for (μ,σ) can easily be constructed.

Christopeit (1994) showed that the method of moments provides consistent estimates of the parameters of extreme value distributions, and used it for estimation of the distribution of earthquake magnitudes in the middle Rhein region. The method of mixed moments (MIX) uses the first moment of the type 1 distribution and the first moment of its logarithmic version. (See Jain and Singh (1987).)

1.7.2 *Simple Linear Estimation*

Noting that the likelihood equations for μ and σ do not admit explicit solutions and hence need to be solved by numerical iterative methods, Kimball (1956) suggested a simple modification to the equation for σ (based on the equation for μ) that makes it easier to solve the resulting equation. The equation for σ given by

$$\hat{\sigma} = \bar{X} - \frac{\sum_{i=1}^{n} X_i e^{-X_i/\hat{\sigma}}}{\sum_{i=1}^{n} e^{-X_i/\hat{\sigma}}} \tag{1.68}$$

used in conjunction with the equation for μ given by

$$\hat{\mu} = -\hat{\sigma}\log\left\{\frac{1}{n}\sum_{i=1}^{n} e^{-X_i/\hat{\sigma}}\right\} \tag{1.69}$$

can be rewritten as

$$\begin{aligned}\hat{\sigma} &= \bar{X} - \frac{1}{n}\sum_{i=1}^{n} X_i e^{-(X_i-\hat{\mu})/\hat{\sigma}}\\ &= \bar{X} + \frac{1}{n}\sum_{i=1}^{n} X_i \log \hat{F}_X(X_i)\,,\end{aligned} \tag{1.70}$$

where as above $\hat{F}_X(X_i)$ is the estimated cumulative distribution function. Replacing $\log \hat{F}_X(X_i')$ in (1.70) with the expected value of $\log \hat{F}_X(X_i')$, Kimball (1956) derived a simplified linear estimator for μ as

$$\hat{\sigma}^* = \bar{X} + \frac{1}{n}\sum_{i=1}^{n} X_i'\left(\sum_{j=1}^{n}\frac{1}{j}\right) \tag{1.71}$$

which may be approximated as

$$\hat{\sigma}^* = \bar{X} + \sum_{i=1}^{n} X_i' \log\left(\frac{i-\frac{1}{2}}{n+\frac{1}{2}}\right). \tag{1.72}$$

The estimator in (1.71) or in (1.72) is a linear function of the order statistics, and hence its bias and mean square error can be determined easily from means, variances and covariances of order statistics. Since the linear estimator in (1.72) is biased, Kimball (1956) presented a table of corrective multipliers; from the table it appears that for $n \geq 10$ the estimator

$$\hat{\sigma}^*(1 + 2.3n^{-1})^{-1} \tag{1.73}$$

is very nearly unbiased. Furthermore a simplified linear estimator of μ may then be obtained as:

$$\text{Estimator of } \mu = \bar{X} - \gamma \times (\text{Estimator of } \sigma). \tag{1.74}$$

Due to the linearity of the estimator of σ, it is only natural to compare it with the best linear unbiased estimator of σ and with its approximations proposed by Blom (1958) and Weiss (1961). (See Tables below.)

Table 1.2: Efficiencies (%) of linear unbiased estimators of μ for the extreme value distribution.

n	2	3	5	6	∞
Best linear	84.05	91.73	95.82	96.65	100.00
Blom's approximation	84.05	91.72	95.68	96.45	100.00
Weiss's approximation	84.05	91.73	95.82	96.63	–
Kimball's approximation	84.05	91.71	95.82	96.63	–

Table 1.3: Efficiencies (%) of linear unbiased estimators of σ for the extreme value distribution.

n	2	3	5	6	∞
Best linear	42.70	58.79	72.96	76.78	100.00
Blom's approximation	42.70	57.47	70.47	74.07	100.00
Weiss's approximation	42.70	58.00	71.04	74.47	–
Kimball's approximation	42.70	57.32	69.88	73.25	–

The location parameter μ can be estimated with quite good accuracy using simple linear functions of order statistics; however, the situation is unsatisfactory should one use such simple linear functions of order statistics to estimate the scale parameter σ. See Tables 1.2 and 1.3.

For the case of Type II right-censored sample $X'_i, X'_2, \ldots, X'_{n-s}$ from the type 1 extreme value distribution for minima with cdf $F_X(x) = 1 - e^{-e^{(x-\mu)/\sigma}}$, Bain (1972) suggested a simple unbiased linear estimator for the scale parameter σ. This estimator was subsequently modified by Engelhardt and Bain (1973) to the form

$$\hat{\hat{\sigma}} = \frac{1}{nk_{n-s,n}} \sum_{i=1}^{n-s} |X'_r - X'_i| , \tag{1.75}$$

where

$$k_{n-s,n} = \frac{1}{n} \sum_{i=1}^{n-s} E|Y'_r - Y'_i| , \tag{1.76}$$

and $Y'_i = (X'_i - \mu)/\sigma$ are the order statistics from the standard type 1 extreme value distribution for minima, while

$$r = \begin{cases} n-s & \text{for } n-s \leq 0.9n , \\ n & \text{for } n-s = n,\ n \leq 15 , \\ n-1 & \text{for } n-s = n,\ 16 \leq n \leq 24 , \\ [0.892n]+1 & \text{for } n-s = n,\ n \geq 25 . \end{cases}$$

Bain (1972) determined exact values of $k_{n-s,n}$ for $n = 5$, 15, 20, 30, 60 and 100 and n infinite, and $(n-s)/n = 0.1(0.1)0.9$ for integer $n-s$. Engelhardt and Bain (1973) gave exact values of $k_{n,n}$ for $n = 2(1)35(5)100$, $n = 39$, 49 and 59 and n infinite. Mann and Fertig (1975) also provided exact values of $k_{n-s,n}$ for $n = 25(5)60$ and $(n-s)/n = 0.1(0.1)1.0$ for integer $n-s$.

Since σ is a scale parameter and $\hat{\hat{\sigma}}$ is an unbiased estimator of σ, improvement is possible in terms of minimum mean-square-error estimator. The improvement in efficiency becomes considerable when the censoring is heavy.

From the tables of Bain (1972) and Engelhardt and Bain (1973), it is evident that the estimator $\hat{\hat{\sigma}}$ in (1.75) is highly efficient; for example when $(n-s)/n \leq 0.7$, the asymptotic efficiency of $\hat{\hat{\sigma}}$ relative to the Cramér–Rao lower bound is at least 97.7%.

The estimator $\hat{\hat{\sigma}}$ in (1.75) may also be used to produce a simple linear unbiased estimator for μ, via the moment equation

$$X'_i = E(X'_i) = \mu + \sigma E(Y'_i) , \tag{1.77}$$

as

$$\hat{\hat{\mu}} = X'_i - E(Y'_i)\hat{\hat{\sigma}} . \tag{1.78}$$

Using the estimators $\hat{\hat{\sigma}}$ and $\hat{\hat{\mu}}$ in Eqs. (1.75) and (1.78), respectively, a simple linear unbiased estimator for the pth quantile μ_p can be derived as

$$\hat{\hat{\mu}}_p = \hat{\hat{\mu}} + \hat{\hat{\sigma}} \log(-\log(1-p)) , \qquad 0 < p < 1 . \tag{1.79}$$

Confidence intervals for the parameters μ and σ based on the linear unbiased estimators $\hat{\hat{\mu}}$ and $\hat{\hat{\sigma}}$ have also been discussed by Bain (1972) and Mann and Fertig (1975).

As pertinently pointed out by Mann and Fertig (1975), for $n - s \leq 0.90n$,

$$\frac{\hat{\hat{\sigma}}}{\sigma} = \frac{1}{nk_{n-s,n}} \sum_{i=1}^{n-s} \frac{X'_{n-s} - X'_i}{\sigma}$$

is approximately a sum of weighted independent chi-square variables.

Thomas and Wilson (1972) also investigated point estimation for the scale and location parameters of the extreme-value (type 1) distribution by linear functions of order statistics from Type II progressively censored samples.

1.7.3 *Best Linear Unbiased (Invariant) Estimation (BLUE)*

Let $X'_{r+1} \leq X'_{r+2} \leq \cdots \leq X'_{n-s}$ be the available doubly Type-II censored sample from a sample of size n where the smallest r and the largest s observations have been censored. Denote

$$\mathbf{X} = (X'_{r+1}, X'_{r+2}, \cdots, X'_{n-s})^{\mathrm{T}},$$

$$\mathbf{1} = (1, 1, \cdots, 1)^{\mathrm{T}}_{1\times(n-r-s)},$$

$$\boldsymbol{\mu} = (E[Y'_{r+1}], E[Y'_{r+2}], \ldots, E[Y'_{n-s}])^{\mathrm{T}}$$

and

$$\boldsymbol{\Sigma} = ((\mathrm{cov}(Y'_i, Y'_j))), \qquad r+1 \leq i, j \leq n-s.$$

Minimizing the generalized variance

$$(\mathbf{X} - \mu\mathbf{1} - \sigma\boldsymbol{\mu})^{\mathrm{T}}\boldsymbol{\Sigma}^{-1}(\mathbf{X} - \mu\mathbf{1} - \sigma\boldsymbol{\mu}),$$

we derive the best linear unbiased estimators (BLUEs) of μ and σ as [see e.g. Balakrishnan and Cohen (1991, pp. 80–81)]:

$$\mu^* = \left\{\frac{\boldsymbol{\mu}^{\mathrm{T}}\boldsymbol{\Sigma}^{-1}\boldsymbol{\mu}\mathbf{1}^{\mathrm{T}}\boldsymbol{\Sigma}^{-1} - \boldsymbol{\mu}^{\mathrm{T}}\boldsymbol{\Sigma}^{-1}\mathbf{1}\boldsymbol{\mu}^{\mathrm{T}}\boldsymbol{\Sigma}^{-1}}{(\boldsymbol{\mu}^{\mathrm{T}}\boldsymbol{\Sigma}^{-1}\boldsymbol{\mu})(\mathbf{1}^{\mathrm{T}}\boldsymbol{\Sigma}^{-1}\mathbf{1}) - (\boldsymbol{\mu}^{\mathrm{T}}\boldsymbol{\Sigma}^{-1}\mathbf{1})^2}\right\}\mathbf{X}$$

$$= \sum_{i=r+1}^{n-s} a_i X'_i \tag{1.80}$$

and

$$\sigma^* = \left\{\frac{\mathbf{1}^{\mathrm{T}}\boldsymbol{\Sigma}^{-1}\mathbf{1}\boldsymbol{\mu}^{\mathrm{T}}\boldsymbol{\Sigma}^{-1} - \mathbf{1}^{\mathrm{T}}\boldsymbol{\Sigma}^{-1}\boldsymbol{\mu}\mathbf{1}^{\mathrm{T}}\boldsymbol{\Sigma}^{-1}}{(\boldsymbol{\mu}^{\mathrm{T}}\boldsymbol{\Sigma}^{-1}\boldsymbol{\mu})(\mathbf{1}^{\mathrm{T}}\boldsymbol{\Sigma}^{-1}\mathbf{1}) - (\boldsymbol{\mu}^{\mathrm{T}}\boldsymbol{\Sigma}^{-1}\mathbf{1})^2}\right\}\mathbf{X}$$

$$= \sum_{i=r+1}^{n-s} b_i X'_i. \tag{1.81}$$

In Table 1.4 the coefficients a_i and b_i are presented for $n = 2(1)\,7$.

Observing that these estimators are minimum variance estimators in the class of all linear unbiased estimators, Mann (1969) considered the larger class of all linear estimators and derived improved estimators by minimizing the mean square error. These estimators are termed the best linear invariant estimators (BLIEs) by Mann (1969); they are particularly useful when either the sample size is very small or there is a substantial censoring in the sample.

Table 1.4: Coefficients for the BLUEs of μ and σ for complete samples. (Balakrishnan and Cohen (1991))

n	i	a_i	b_i
2	1	0.91637	−0.72135
2	2	0.08363	0.72135
3	1	0.65632	−0.63054
3	2	0.25571	0.25582
3	3	0.08797	0.37473
4	1	0.51100	−0.55862
4	2	0.26394	0.08590
4	3	0.15368	0.22392
4	4	0.07138	0.24880
5	1	0.41893	−0.50313
5	2	0.24628	0.00653
5	3	0.16761	0.13045
5	4	0.10882	0.18166
5	5	0.05835	0.18448
6	1	0.35545	−0.45927
6	2	0.22549	−0.03599
6	3	0.16562	0.07320
6	4	0.12105	0.12672
6	5	0.08352	0.14953
6	6	0.04887	0.14581
7	1	0.30901	−0.42370
7	2	0.20626	−0.06070
7	3	0.15859	0.03619
7	4	0.12322	0.08734
7	5	0.09375	0.11487
7	6	0.06733	0.12586
7	7	0.04184	0.12014

Denoting the BLIEs of μ and σ by

$$\mu^{**} = \sum_{i=r+1}^{n-s} a_i^* X_i' \qquad \text{and} \qquad \sigma^{**} = \sum_{i=r+1}^{n-s} b_i^* X_i' . \tag{1.82}$$

Mann (1967a) in a Technical Report and Mann, Schafer and Singpurwalla (1974) in the by now classical volume on Reliability have presented tables for various sample sizes and different levels of censoring. In Table 1.5, the coefficients a_i^* and b_i^* are presented for $n = 2(1)7$ for the case of complete samples (i.e. $r = s = 0$).

Analysis of BLIEs reveals that while there is only a slight improvement in the estimation of μ, there is a significant gain in using the BLIE of σ, particularly when n is small.

Table 1.5: Coefficients for the BLIEs of μ and σ for complete samples. (Mann, Schafer and Singpurwalla (1974))

n	i	a_i^*	b_i^*
2	1	0.88927	−0.42138
2	2	0.11073	0.42138
3	1	0.66794	−0.46890
3	2	0.25100	0.19024
3	3	0.08106	0.27867
4	1	0.52681	−0.45591
4	2	0.26151	0.07011
4	3	0.14734	0.18275
4	4	0.06434	0.20305
5	1	0.43359	−0.43126
5	2	0.24609	0.00560
5	3	0.16381	0.11182
5	4	0.10353	0.15571
5	5	0.05298	0.15813
6	1	0.36818	−0.40573
6	2	0.22649	−0.03180
6	3	0.16359	0.06467
6	4	0.11754	0.11195
6	5	0.07938	0.13210
6	6	0.04483	0.12881
7	1	0.31993	−0.38202
7	2	0.20783	−0.05472
7	3	0.15766	−0.03263
7	4	0.12097	0.07875
7	5	0.09079	0.10357
7	6	0.06409	0.11348
7	7	0.03874	0.10832

1.7.4 *Asymptotic Best Linear Unbiased Estimation*

Optimal linear estimation of the parameters μ and σ based on k selected order statistics, using the theory of Ogawa (1951, 1952), has been discussed by a number of authors. Suppose that $0 < \lambda_1 < \lambda_2 < \cdots < \lambda_k < 1$ is the spacing that needs to be determined optimally, and let $\lambda_0 = 0$ and $\lambda_{k+1} = 1$. $X'_{n_i:n}$ is called the sample quantile of order λ_i, where $n_i = [n\lambda_i] + 1$. Then it can be shown that the asymptotic variances and covariance of the BLUEs, $\tilde{\mu}^*$ and $\tilde{\sigma}^*$, based on the k selected sample quantiles are given by

$$\operatorname{Var}(\tilde{\mu}^*) = \frac{\sigma^2}{n} \cdot \frac{K_{22}}{K_{11}K_{22} - K_{12}^2}, \tag{1.83}$$

$$\operatorname{Var}(\tilde{\sigma}^*) = \frac{\sigma^2}{n} \cdot \frac{K_{11}}{K_{11}K_{22} - K_{12}^2} \tag{1.84}$$

and

$$\mathrm{Cov}(\tilde{\mu}^*, \tilde{\sigma}^*) = -\frac{\sigma^2}{n} \cdot \frac{K_{12}}{K_{11}K_{22} - K_{12}^2}. \tag{1.85}$$

In the equations above

$$K_{11} = \sum_{i=1}^{k+1} \frac{\{p_Y(G_i) - p_Y(G_{i-1})\}^2}{\lambda_i - \lambda_{i-1}}, \tag{1.86}$$

$$K_{12} = \sum_{i=1}^{k+1} \frac{\{p_Y(G_i) - p_Y(G_{i-1})\}\{G_i p_Y(G_i) - G_{i-1}p_Y(G_{i-1})\}}{\lambda_i - \lambda_{i-1}} \tag{1.87}$$

and

$$K_{22} = \sum_{i=1}^{k+1} \frac{\{G_i p_Y(G_i) - G_{i-1}p_Y(G_{i-1})\}^2}{\lambda_i - \lambda_{i-1}}, \tag{1.88}$$

where $G_i = F_Y^{-1}(\lambda_i)$ and the quantities

$$p_Y(G_0)\,, \qquad G_0 p_Y(G_0)\,, \qquad p_Y(G_{k+1})\,, \qquad G_{k+1}p_Y(G_{k+1})$$

vanish.

Appropriate functions involving K_{11}, K_{22} and K_{12} need to be optimized, subject to the constraint $0 < \lambda_1 < \lambda_2 < \cdots < \lambda_k < 1$ in order to determine the k optimal quantiles for the asymptotic BLU estimation of the parameters μ and σ. Numerical results for this problem have been provided by Hassanein (1965, 1968, 1969, 1972) and Chan and Kabir (1969). As an example the optimal spacing $(\lambda_1, \lambda_2, \ldots, \lambda_k)$ that maximizes K_{11} in (1.86) is presented in Table 1.6 for $k = 1(1)7$. These values provide the optimal sample quantiles to be used in a sample of size n for the asymptotic BLU estimator of μ (when σ is known) since its variance in this case is given by

$$\mathrm{Var}(\tilde{\mu}^*) = \frac{\sigma^2}{nK_{11}}. \tag{1.89}$$

Tests of hypotheses about the equality of μ's from several extreme value populations based on asymptotic BLU estimators are discussed by Hassanein and Saleh (1992).

Table 1.6: Optimal spacing for the asymptotic best linear unbiased estimator of μ (when σ is known) for $k = 1(1)7$.

k	λ_1	λ_2	λ_3	λ_4	λ_5	λ_6	λ_7
1	0.2032						
2	0.0734	0.3615					
3	0.0345	0.1701	0.4705				
4	0.0190	0.0933	0.2581	0.5486			
5	0.0115	0.0566	0.1566	0.3329	0.6069		
6	0.0075	0.0369	0.1021	0.2171	0.3958	0.6521	
7	0.0052	0.0254	0.0703	0.1494	0.2723	0.4487	0.6880

1.7.5 *Maximum Likelihood Estimation*

Maximum likelihood estimation of extreme values distributions is a subject to which numerous studies are devoted. This method will be efficient in any of the following cases:

(i) The distribution is of Gumbel type (type 1) with coefficients of location and dispersion unknown;

(ii) The distribution is of Fréchet (type 2) or of Weibull (type 3) type, with known parameter of location (in that case, for a type 2 distribution, $\log(X_i - \lambda)$ has a type 1 distribution, or equivalently $-\log(\lambda - X_i)$ for a type 3 distribution ($x = \lambda$ being in both cases the origin of the distribution), and unknown parameters of shape and dispersion;

(iii) The distribution is of Fréchet (type 2) type, with three parameters (location, shape and dispersion) unknown;

(iv) The distribution is of Weibull (type 3) type, with three parameters (location, shape and dispersion) unknown, k being restricted to be ≥ 2.

The main difficulties thus appear for the Weibull distribution with *unknown* shape parameter.

Below we shall discuss exclusively the maximum likelihood estimation for the type 1 (Gumbel) extreme value distribution.

Gumbel (1958) argued over forty years ago — that the method of maximum likelihood estimation (MLE) was very complicated and required numerical work normally prohibitive in that time for routine use and favored the method of moments (MOM). Lettenmaier and Burges (1982) showed some 25 years later that the MLE method gave better parameter estimates than those by the MOM method, especially for large return periods and small sample sizes.

A Fortran 77-Program GEMPAK developed by Al Abbasi and Fahmi (1991) estimated parameters of type 1, type 3 and the so-called mixture upper earthquake magnitude extremal asymptotic distributions by means of the maximum likelihood method with numerical maximization utilizing the Newton–Raphson procedure. The subprogram calculates return periods at magnitude classes regarded as risky.

Complete Data Case

Based on a random sample $X_1, X_2, \ldots, X_n$, the maximum likelihood estimators μ and σ satisfy the equations

$$\sum_{i=1}^{n} e^{-(X_i - \hat{\mu})/\hat{\sigma}} = n \tag{1.90}$$

and

$$\sum_{i=1}^{n} (X_i - \hat{\mu})(1 - e^{-(X_i - \hat{\mu})/\hat{\sigma}}) = n\hat{\sigma}\,. \tag{1.91}$$

The asymptotic variances of $\hat{\mu}$ and $\hat{\sigma}$ are given by the Cramér–Rao lower bounds in (1.61). The asymptotic correlation coefficient between $\hat{\mu}$ and $\hat{\sigma}$ is

$$\left\{1 + \frac{\pi^2}{\sigma(1-\gamma)^2}\right\}^{-1/2} = 0.313\,. \tag{1.92}$$

Equation (1.90) can be rewritten as

$$\hat{\mu} = -\hat{\sigma} \log \left(\frac{1}{n} \sum_{i=1}^{n} e^{-X_i/\hat{\sigma}} \right) ; \tag{1.93}$$

this, when used in Eq. (1.91), yields the following equation for $\hat{\sigma}$:

$$\hat{\sigma} = \bar{X} - \frac{\sum_{i=1}^{n} X_i e^{-X_i/\hat{\sigma}}}{\sum_{i=1}^{n} e^{-X_i/\hat{\sigma}}} . \tag{1.94}$$

It is necessary to solve (1.94) by an iterative method for $\hat{\sigma}$; Eq. (1.93) will then give $\hat{\mu}$. If $\hat{\sigma}$ is large compared to X_i's, then the right-hand side of (1.94) is approximately

$$\bar{X} \left\{ 1 - \frac{n-1}{n} \cdot \frac{S^2}{\hat{\sigma}\bar{X}} \right\} . \tag{1.95}$$

This will provide an approximate solution to (1.94) which can be used as an initial guess for the iterative method to solve Eq. (1.94).

The asymptotic confidence intervals at significance level α are given by

$$\left(\frac{\hat{\mu} - \mu}{\sigma} \right)^2 - 2(1-\gamma) \left(\frac{\hat{\mu} - \mu}{\sigma} \right) \left(\frac{\hat{\sigma} - \sigma}{\sigma} \right) + \left\{ \frac{\pi^2}{6} + (1-\gamma)^2 \right\} \left(\frac{\hat{\sigma} - \sigma}{\sigma} \right)^2$$
$$\leq -\frac{2}{n} \log \alpha ,$$

i.e.

$$\left(\frac{\hat{\mu} - \mu}{\sigma} \right)^2 - 0.84556 \left(\frac{\hat{\mu} - \mu}{\sigma} \right) \left(\frac{\hat{\sigma} - \sigma}{\sigma} \right) + 1.82367 \left(\frac{\hat{\sigma} - \sigma}{\sigma} \right)^2 \leq -\frac{2}{n} \log \alpha .$$

Evidently, these are ellipses in the (μ, σ) plane. For the estimator

$$\hat{\mu}_p = \hat{\mu} - \log(-\log p)\hat{\sigma}$$

of the pth percentile of the distribution, the asymptotic variance is

$$\frac{\sigma^2}{n} \left[1 + \frac{6}{\pi^2} \{ 1 - \gamma - \log(-\log p) \}^2 \right] .$$

Tiago de Oliveira (1972) has shown that the best asymptotic point predictor of the maximum of (the next) m observations is

$$\hat{\mu} + (\gamma + \log m)\hat{\sigma}$$

with the asymptotic variance

$$\frac{\sigma^2}{n} \left[1 + \frac{6}{\pi^2} \{ 1 + \log m \}^2 \right] .$$

If the scale parameter σ is known, the maximum likelihood estimator of μ is obtained from (1.90) to be

$$\hat{\mu}_1 = -\sigma \log \left\{ \frac{1}{n} \sum_{i=1}^{n} e^{-X_i/\sigma} \right\} . \tag{1.96}$$

This estimator is not unbiased. In fact Kimball (1956) has shown that (when σ is known)

$$E[\hat{\mu}_{1|\sigma}] = \mu + \sigma\left\{\gamma + \log n - 1 - \frac{1}{2} - \cdots - \frac{1}{n-1}\right\} \tag{1.97}$$

and

$$\mathrm{Var}(\hat{\mu}_{1|\sigma}) = \sigma^2\left\{\frac{\pi^2}{6} - 1^2 - \frac{1}{2^2} - \cdots - \frac{1}{(n-1)^2}\right\}. \tag{1.98}$$

While $\hat{\mu}_{1|\sigma}$ is a biased estimator of μ, $e^{-\hat{\mu}_{1|\sigma}/\sigma}$ is an unbiased estimator of $e^{-\mu/\sigma}$. This is so because $e^{-X/\sigma}$ has an exponential distribution with expected value $e^{-\mu/\sigma}$.

Posner (1965), when applying the extreme value theory to error-free communication, estimated the parameters μ and σ for the complete sample case by the maximum likelihood theory and justified its use on the basis of its asymptotic properties. Observing that the asymptotic theory needs not be valid for Posner's sample size ($n = 30$), Gumbel and Mustafi (1966) showed that in fact a modified method of moments gives better results for Posner's data.

Censored Case

Suppose that the available sample is a doubly Type-II censored sample $X'_{r+1}, X'_{r+2}, \ldots, X'_{n-s}$. Then the log-likelihood function based on this censored sample is

$$\begin{aligned}\log L = {} & \log n! - \log r! - \log s! - \sum_{i=r+1}^{n-s} Y'_i - \sum_{i=r+1}^{n-s} e^{-Y'_i} \\ & - (n-r-s)\log\sigma + r\log F_Y(Y'_{r+1}) + s\log\{1 - F_Y(Y'_{n-s})\},\end{aligned} \tag{1.99}$$

where $Y'_i = (X'_i - \mu)/\sigma$ are the order statistics from the standard type 1 extreme value distribution with density (1.22) and $F_Y(y)$ is the corresponding cdf. From (1.99), we obtain the likelihood equations for μ and σ to be

$$\frac{\partial \log L}{\partial \mu} = \frac{1}{\sigma}\left[(n-r-s) - \sum_{i=r+1}^{n-s} e^{-Y'_i} - r\frac{p_Y(Y'_{r+1})}{F_Y(Y'_{r+1})} + s\frac{p_Y(Y'_{n-s})}{1 - F_Y(Y'_{n-s})}\right] = 0 \tag{1.100}$$

and

$$\begin{aligned}\frac{\partial \log L}{\partial \sigma} = \frac{1}{\sigma}\Bigg[& \sum_{i=r+1}^{n-s} Y'_i - \sum_{i=r+1}^{n-s} Y'_i e^{-Y'_i} - (n-r-s) \\ & - rY'_{r+1}\frac{p_Y(Y'_{r+1})}{F_Y(Y'_{r+1})} + sY'_{n-s}\frac{p_Y(Y'_{n-s})}{1 - F_Y(Y'_{n-s})}\Bigg] = 0.\end{aligned} \tag{1.101}$$

Harter and Moore (1968a) and Harter (1970) have discussed the numerical solution of the above likelihood equations. The asymptotic variance–covariance matrix of the maximum likelihood estimates, $\hat{\mu}$ and $\hat{\sigma}$, determined from Eqs. (1.100) and (1.101) is given by [Harter (1970, pp. 127–128)]:

$$\frac{\sigma^2}{n}\begin{bmatrix} V_{11} & V_{12} \\ V_{12} & V_{22} \end{bmatrix}, \tag{1.102}$$

where $((V_{ij}))$ is the inverse of the matrix $((V^{ij}))$ with

$$V^{11} = 1 - q_1 - q_2 + q_1 \log q_1 - (1 - q_2) \log (1 - q_2) ,$$

$$\begin{aligned} V^{22} = &-(1 - q_1 - q_2) - 2\{\Gamma'(1; -\log q_1) - \Gamma'(1; -\log (1 - q_2))\} \\ &- \Gamma''(2; -\log q_1) - \Gamma''(2; -\log (1 - q_2)) + 2\{\Gamma'(2; -\log q_1) \\ &- \Gamma'(2; -\log (1 - q_2))\} - q_1 \log q_1 \log (-\log q_1) \\ &\times \{2 + \log (-\log q_1)\} + (1 - q_2) \log (1 - q_2) \log \{-\log (1 - q_2)\} \\ &\times \left[2 + \log \{-\log (1 - q_2)\} + \log (1 - q_2) \frac{\log \{-\log (1 - q_2)\}}{q_2}\right] \end{aligned}$$

and

$$\begin{aligned} V^{12} &= V^{21} \\ &= -\Gamma'(2; -\log q_1) + \Gamma'(2; -\log (1 - q_2)) + q_1 \log (q_1) \log (-\log q_1) \\ &\quad - (1 - q_2) \log (1 - q_2) \log \{-\log (1 - q_2)\} \\ &\quad - \left(\frac{1}{q_2} - 1\right) \log^2 (1 - q_2) \log \{-\log (1 - q_2)\} . \end{aligned}$$

In the equations above:
$q_1 = r/n$, $q_2 = s/n$, $\Gamma(p; a) = \int_0^a e^{-t} t^{p-1} \, dt$, $\Gamma'(p; a) = (d/du)\Gamma(u; a)_{u=p}$.

Harter (1970) has tabulated the values of V_{11}, V_{12} and V_{22} for $q_1 = 0.0(0.1)0.9$ and $q_2 = 0.0(0.1)(0.9 - q_1)$.

Phien (1991) has discussed further the maximum likelihood estimation of the parameters μ and σ based on censored samples. He carried out an extensive simulation study and observed the following concerning the effects of Type I censoring on the estimation of parameters and quantiles of the type 1 extreme value distribution using the maximum likelihood method: (a) light censoring on the right may be useful in reducing the bias in estimating the parameters; (b) the bias in estimating the parameters and quantiles is very small; (c) for complete samples the MLE of μ overestimates μ, while the MLE of σ underestimates σ slightly; and (d) censoring introduces an increase in the variances of the estimates.

For the distribution

$$F_X(x) = e^{-e-(x-\mu)/\sigma} ,$$

with X_l and X_r as the left- and right-censoring time points and with r lowest and s largest observations censored (doubly Type-I censored data), the likelihood function is proportional

$$\{F_X(X_l)\}^r \prod_{i=r+1}^{n-s} p_X(X_i) \{1 - F_X(X_r)\}^s . \qquad (1.101a)$$

In this case that r and s are random variables while X_l and X_r are fixed. The log-likelihood function is

$$\log L = \text{const} - (n - r - s) \log \sigma - \sum_{i=r+1}^{n-s} \{Y_i + e^{-Y_i}\} - rd + s \log q .$$

Here

$$d = e^{-Y_l},$$

$$q = 1 - e^{-e^{-Y_r}},$$

$$Y_l = \frac{X_l - \mu}{\sigma}, \qquad \text{and similarly for } Y_r \text{ and } Y_i.$$

The maximum likelihood estimators of μ and σ satisfy the equations

$$\frac{\partial \log L}{\partial \sigma} = -\frac{G}{\sigma} = 0 \qquad \text{and} \qquad \frac{\partial \log L}{\partial \mu} = -\frac{H}{\sigma} = 0,$$

where

$$G = P + P_l + P_r \qquad \text{and} \qquad H = Q + Q_l + Q_r$$

with

$$P = n - r - s - \sum_{i=r+1}^{n-s} Y_i + \sum_{i=r+1}^{n-s} Y_i e^{-Y_i}, \qquad Q = -(n - r - s) + \sum_{i=r+1}^{n-s} e^{-Y_i},$$

$$P_l = rdY_l,$$

$$Q_l = rd,$$

$$P_r = \frac{se^{Y_r}(1-q)Y_r}{q},$$

$$Q_r = \frac{se^{-Y_r}(1-q)}{q}.$$

Phien (1991) recommended solving these equations using Newton's iterative process. Simulations carried out by Bugaighis (1991) show the ML estimator for σ to have a slight edge over the BLU, particularly for very small ($n < 10$) samples and heavy censorship. However, this slight advantage of the ML estimator of σ dissipates with increasing the sample size. This is particularly noticeable in the case of moderate to light censorship. The situation is reversed when it comes to estimating the location parameter μ. In this case, the BLU estimator of μ is the more efficient of the two. Recall that similar results were reported by Mann *et al.* (1974), when considering moderate forms of Type II censorship. (Type II censorship is considered moderate when, in reliability terminology, at least 50% of the tested items are actually observed to fail.) Evidently, further investigations are desirable.

An alternative approach was taken by Balakrishnan and Varadan (1991), who approximated the likelihood equations by using appropriate linear functions and derived *approximate* maximum likelihood estimators of μ and σ. They derived these estimators for the type 1 extreme value distribution for the minimum. [The estimators for the type 1 extreme value distribution for the maximum in (1.21) can be obtained simply by interchanging r and s and replacing μ by $-\mu$ and X_i' by $-X_{n-i+1}'$.] A simulation study, Balakrishnan and Varadan (1991), demonstrates that their estimators are as efficient as the maximum likelihood estimators, BLU estimators, and BLI estimators (even for samples of size as small as 10).

Estimators of this type based on multiply Type-II censored samples have also been discussed by Balakrishnan, Gupta, and Panchapakesan (1992) and Fei, Kong, and Tang (1994), among others.

1.7.6 *Method of Probability-Weighted Moments* (*PWM*)

Another method popular in extreme value investigations (especially in environmental sciences) is the PWM method. Landwehr, Matalas and Wallis (1979) proposed this method of estimation of the parameters μ and σ based on probability-weighted moments defined as:

$$M_{(k)} = E[X\{1 - F(X)\}^k], \qquad k = 0, 1, 2, \dots .$$

An unbiased estimator of $M_{(k)}$ is given by

$$\hat{M}_{(k)} = \frac{1}{n} \sum_{i=1}^{n} X_i' \frac{\binom{n-1}{k}}{\binom{n-1}{k}}, \qquad k = 0, 1, 2, \dots .$$

By making use of the explicit expressions of $M_{(0)}$ and $M_{(1)}$, equating them to sample estimators $\hat{M}_{(0)}$ and $\hat{M}_{(1)}$ and solving for the parameters μ and σ, these authors derived the probability-weighted moments estimators to be

$$\hat{\sigma} = \frac{\hat{M}_{(0)} - 2\hat{M}_{(1)}}{\log 2} \qquad \text{and} \qquad \hat{\mu} = \hat{M}_{(0)} - \gamma\hat{\sigma} .$$

They compared the performance of these estimators with the moment estimators and the maximum likelihood estimators in terms of bias and the mean square error. Their extensive simulation study indicated that this method of estimation is simple and also highly efficient (in terms of the efficiency relative to the maximum likelihood estimates). See Table 1.7.

Table 1.7: Bias, mean square error and relative efficiency of the moment estimators, PWM estimators, and ML estimators μ and σ based on a complete sample of size n. (Landwehr *et al.* (1979))

		σ			μ		
Method	n	Bias	MSE	Relative Efficiency	Bias	MSE	Relative Efficiency
M	5	0.18	0.37	0.83	−0.10	0.49	0.97
PWM		0.15	0.34	1	−0.08	0.49	1
ML		0.00	0.44	0.74	0.01	0.48	1.05
M	9	0.11	0.30	0.74	−0.06	0.36	0.96
PWM		0.09	0.26	1	−0.04	0.36	1
M	49	0.02	0.14	0.60	−0.01	0.15	0.96
PWM		0.02	0.11	1	0.00	0.15	1
ML		0.00	0.13	0.77	0.00	0.15	1.00

More details on this method, its drawbacks and advantages, are given in Chap. 2, in the section dealing with estimation of parameters in the case of generalized extreme value distributions.

1.7.7 *Ranked Set Estimation*

Bhoj (1997), Barnett and Moore (1997) and Barnett (1999) investigated ranked set sample design for various distributions including the type 1 extreme value distribution with the cdf

$$G_X\left(\frac{x-\mu}{\sigma}\right) = \exp\left[-\exp\left[-\left(\frac{x-\mu}{\sigma}\right)\right]\right].$$

Consider $X_{(1)}, X_{(2)}, \ldots, X_{(n)}$, the order statistics of a sample of size n, and let $U_{(i)} = (X_{(i)} - \mu)/\sigma$. Define $\alpha_i = E(U_{(i)})$ and $v_i = \mathrm{Var}(U_{(i)})$. Suppose the ranked set sample $x_{1(1)}, x_{2(2)}, \ldots, x_{n(n)}$ is obtained as the set of smallest, second smallest, up to largest, observed values in n conceptual samples $x_{i1}, x_{i2}, \ldots, x_{in}$ $(i = 1, 2, \ldots, n)$ under the assumption that correct ordering has taken place (*incorrect* ordering can also be allowed for; see Barnett and Moore (1997)). We shall use the ranked set sample for estimation of μ and σ. The usual estimator of $E(X)$ is the *ranked set sample mean*

$$\overline{\overline{X}} = \frac{1}{n}\sum_{1}^{n} X_{i(i)} \tag{1.102a}$$

which is known to be unbiased, with variance $\sigma^2 \sum v_i/n^2$.

We have:

$$\mathrm{Var}(\overline{\overline{X}}) = \frac{\sigma_X^2}{n} - \sigma^2 \sum (\alpha_i - \bar{\alpha})^2/n^2$$

(which confirms the fact that the ranked set sample mean $\overline{\overline{X}}$, cannot be less efficient than the sample mean, $\bar{X}$). The *relative efficiency* is then:

$$e(\overline{\overline{X}}, \bar{X}) = \mathrm{Var}(\bar{X})/\mathrm{Var}(\overline{\overline{X}}) = \left\{1 - \sum(\alpha_i - \bar{\alpha})^2\sigma^2/n\sigma_X^2\right\}^{-1}. \tag{1.103}$$

There is no reason why, as in (1.102a), we should adopt *equal weights* for each $X_{i(i)}$. *Optimally chosen weights* should (by definition) provide a gain in efficiency of estimation of $E(X)$. Barnett and Moore (1997) obtained the *ranked set best linear unbiased estimators* (ranked set BLUEs) of μ and σ in $G\{(x-\mu)/\sigma\}$ and hence of $E(X)$. The BLUEs of μ and σ are of the form:

$$\mu^* = \sum \gamma_i X_{i(i)}$$

$$\sigma^* = \sum \eta_i X_{i(i)} \qquad \text{for some } \gamma_i \text{ and } \eta_i \text{ which depend on } \alpha_i \text{ and } v_i\,.$$

For the type 1 extreme value distribution, the reduced variable $U = (X - \mu)/\sigma$ has the mean γ (Euler's constant) and variance $\pi^2/6$. Hence

$$\mu_X = \mu + \gamma\sigma$$

$$\sigma_X^2 = \pi^2\sigma^2/6 \qquad \text{(cf. Sec. 1.2)}.$$

Barnett (1999) found, inter alia, that the relative efficiencies $e(\overline{\overline{X}}, \bar{X})$ in the case of the Gumbel type 1 distributions are as follows:

n	2	3	4	5	6	8	10	15	20
$e(\overline{\overline{X}}, \bar{X})$	1.41	1.79	2.15	2.50	2.83	3.47	4.08	5.53	9.59

The efficiency gains are thus 150%, 250%, 450% and 860% for $n = 5$, 10, 15 and 20 respectively!

1.7.8 *Conditional Method*

The conditional method of inference for location and scale parameters, first suggested by Fisher (1934) and discussed in detail by Lawless (1982), has been used effectively for the type 1 extreme value distribution by Lawless (1973, 1978) and Viveros and Balakrishnan (1994). These developments are described here for the type 1 extreme value distribution for minimum with the cdf $1 - e^{-e^{(x-\mu)/\sigma}}$.

Suppose that $X_1' \leq X_2' \leq \cdots \leq X_{n-s}'$ is the available Type-II right-censored sample. The joint density function of $\mathbf{X} = (X_1', X_2', \ldots, X_{n-s}')$ is

$$p_X(x;\mu,\sigma) = \frac{n!}{s!\sigma^{n-s}} \prod_{i=1}^{n-s} p_Y\left(\frac{x_i'-\mu}{\sigma}\right)\left\{1 - F_Y\left(\frac{x_i'-\mu}{\sigma}\right)\right\}^s, \tag{1.104}$$

where $F(\cdot)$ and $p(\cdot)$ are the cdf and pdf of the standard form of the type 1 extreme value distribution for minimum given by

$$F_Y(y) = 1 - e^{-e^y} \qquad \text{and} \qquad p_Y(y) = e^y e^{-e^y}. \tag{1.105}$$

The joint density in (1.104) preserves the location-scale structure since from (1.104) the standardized variables, $(X_1' - \mu)/\sigma, \ldots, (X_{n-s}' - \mu)/\sigma$ have a joint distribution functionally independent of μ and σ. Suppose that $\hat{\mu}$ and $\hat{\theta}$ are the maximum likelihood estimates of μ and σ (or some equivariant estimators like BLUEs or BLIEs) which jointly maximize the likelihood of (μ, σ) that is proportional to (1.104). Then, $Z_1 = (\hat{\mu} - \mu)/\sigma$ and $Z_2 = \hat{\sigma}/\sigma$ are the pivotal quantities so that their joint density involves neither μ nor σ. With $A_i = (X_i' - \hat{\mu})/\hat{\sigma}$ $(i = 1, 2, \ldots, n-s)$, $\mathbf{A} = (A_1, A_2, \ldots, A_{n-s})$ forms an ancillary statistic, and inferences for μ and σ may thus be based on the joint distribution of Z_1 and Z_2 conditional on the observed value $\mathbf{a}$ of $\mathbf{A}$.

Using $p(z_1, z_2|a)$, Lawless (1973, 1978) applied algebraic manipulations and numerical integration techniques to determine the marginal conditional densities $p(z_1|a)$ and $p(z_2|a)$ that can be used to carry out individual inferences on the parameters.

Viveros and Balakrishnan (1994) have developed a similar conditional method of inference based on Type-II *progressively* censored data when one or more surviving items may be removed from the life-test (or progressively censored) at the time of each failure occurring prior to the termination of the experiment. The complete sample case or the Type-II right-censored sample case are, of course, special cases of this scheme.

1.7.9 *Tolerance Limits*

Dasgupta and Bhaumik (1995) proposed the following direct approach to construction of tolerance limits for extreme value distributions.

(1) Consider type 3 distribution:

$$F(x) = \begin{cases} \exp\left(-\left(\frac{x-\mu}{\sigma}\right)\right)^{\xi}, & x < \mu; \quad \xi \text{ and } \sigma > 0 \\ 1 & x \geq \mu \end{cases}$$

(a negatively skewed distribution)

It is required to determine $x_{(1)}\delta(\mu, \xi, \sigma)$, where δ is a positive function of μ, ξ and σ, and $x_{(1)} = \min(x_1, \ldots, x_n)$, where x_i's are i.i.d. with distribution F, such that

$$P_{x_{(1)}}[P_F\{Y \geq x_{(1)}\delta(\mu, \xi, \sigma)\} \geq \beta] = \gamma$$

for preassigned probabilities β and γ. (Here Y is a future observation from F and we search for a lower bound such that $100\beta\%$ of the future observations will be above that bound with a very high probability γ.) Equivalently,

$$P[1 - F(x_{(1)}\delta(\mu, \xi, \sigma)) \geq \beta] = \gamma$$

or

$$P[x_{(1)}\delta(\mu, \xi, \sigma) \leq F^{-1}(\bar{\beta})] = \gamma \qquad (\text{where } \bar{\beta} = 1 - \beta)$$

or

$$P[x_{(1)} > F^{-1}(\bar{\beta})/\delta(\mu, \xi, \sigma)] = 1 - \gamma\,.$$

Denoting $G = 1 - F$, we obtain

$$G^n[F^{-1}(\bar{\beta})/\delta(\mu, \xi, \sigma)] = 1 - \gamma$$

or

$$F[F^{-1}(\bar{\beta})/\delta(\mu, \xi, \sigma)] = 1 - (1 - \gamma)^{1/n}\,.$$

Thus

$$\delta(\mu, \xi, \sigma) = F^{-1}(\bar{\beta})/F^{-1}(1 - (1 - \gamma)^{1/n})\,.$$

Since for the type 3 distribution

$$F^{-1}(y) = x = \mu - \sigma(-\log y)^{1/\xi}$$

we have

$$\delta(\mu, \xi, \sigma) = \frac{\mu - \sigma(-\log \bar{\beta})^{1/\xi}}{\mu - \sigma[-\log\{1 - (1 - \gamma)^{1/n}\}]^{1/\xi}}\,.$$

The next step is, of course, to estimate the parameters μ, σ and ξ.

2) Similar arguments for the type 1 distribution (positively skewed) with the infinite range:

$$F(x) = \exp\left[-e^{-(\frac{x-\mu}{\sigma})}\right], \qquad -\infty < x < \infty\,, \sigma > 0\,, \mu \in R\,,$$

show that

$$F^{-1}(y) = \mu - \sigma \log(-\log y) \qquad \text{and}$$

$$\delta(\mu, \sigma) = F^{-1}(\bar{\beta})/F^{-1}(1 - (1 - \gamma)^{1/n})\,;$$

thus, in this case,

$$\delta(\mu,\sigma) = \frac{\mu - \sigma \log(-\log \bar{\beta})}{\mu - \sigma \log\{-\log(1-(1-\gamma)^{1/n})\}},$$

and $x_{(1)}\delta(\mu^*, \sigma^*)$ serves as an approximate *lower* tolerance limit where μ^* and σ^* are some consistent estimators of μ and σ.

3) For a type 2 distribution

$$F(x) = \begin{cases} 0\,, & x < \mu \\ \exp\left(-\left(\frac{x-\mu}{\sigma}\right)^{-\frac{1}{\xi}}\right), & x \geq \mu\,, \qquad \xi > 0\,, \sigma > 0\,, \end{cases}$$

there exists no finite lower bound for the variable. Here an *upper* β-content tolerance limit is required. One requires an upper bound such that a large percentage of future observations will be below that bound with high probability (for example, an excessive concentration of ozone causes rise in global temperature — the so-called "greenhouse effect"). We thus consider upper tolerance of the type

$$x_{(n)}\delta\,,$$

where $x_{(n)} = \max(x_1, \ldots, x_n)$ and $\delta > 0$. We need to have

$$P_{x_{(n)}}[P_F\{Y \leq x_{(n)}\delta\} \geq \beta] = \gamma\,,$$

where, as before, Y is a future observation from F; namely at least $100\beta\%$ of the future observations would be below $x_{(n)}\delta$ with a high probability γ.

Similar arguments show that

$$P_{x_{(n)}}[F(x_{(n)}\delta) \geq \beta] = \gamma$$

or

$$P_{x_{(n)}}[x_{(n)} < F^{-1}(\beta)/\delta] = 1 - \gamma$$

and

$$\delta(\mu, \alpha, \sigma) = F^{-1}(\beta)/F^{-1}[(1-\gamma)^{1/n}]\,. \tag{1.106}$$

Since for a type 2 distribution

$$F^{-1}(y) = y + \sigma(-\log y)^{-\xi}$$

and

$$\delta(\mu,\xi,\sigma) = \frac{\mu + \sigma(-\log\beta)^{-\xi}}{\mu + \sigma[-\frac{1}{n}\log((1-\gamma)]^{-\xi}}, \tag{1.107}$$

$x_{(n)}\delta(\mu^*, \xi^*, \sigma^*)$ is an approximate upper tolerance limit where μ^*, ξ^* and σ^* are sample estimates of μ, ξ and σ respectively.

Observe that for the type 1 distribution we also have for the upper tolerance limit:

$$\delta(\mu,\sigma) = \frac{\mu - \sigma\log(-\log\beta)}{\mu - \sigma\log(-\frac{1}{n}\log(1-\gamma))}$$

and the approximation is

$$x_{(n)}\delta(\mu^*, \sigma^*)\,.$$

A slightly different, more flexible approach, popular in engineering applications (using a more common notation) is as follows: based on a complete sample (or Type-II censored sample) observed from the distribution, the lower α tolerance limit for proportion $1-\gamma$ is $\hat{\mu} + k_L\hat{\sigma}$ satisfying the equation

$$\Pr[\Pr[X \geq \hat{\mu} + k_L\hat{\sigma}] \geq 1-\gamma] = \alpha\,; \tag{1.108}$$

similarly the upper α tolerance limit for proportion $1-\gamma$ is $\hat{\mu} + k_U\hat{\sigma}$ satisfying the equation

$$\Pr[\Pr[X \leq \hat{\mu} + k_U\hat{\sigma}] \geq 1-\gamma] = \alpha\,. \tag{1.109}$$

The constants k_L and k_U are referred to as the lower and upper tolerance factors respectively.

In the case of the type 1 extreme value distribution for the minima with the cdf

$$F_X(x) = 1 - e^{-e^{(x-\mu)/\sigma}}\,,$$

Eqs. (1.108) and (1.109) become

$$\Pr\left[\frac{\hat{\mu}-\mu}{\sigma} + k_L\frac{\hat{\sigma}}{\sigma} \leq \log\left[-\log\left(1-\gamma\right)\right]\right] = \alpha \tag{1.110}$$

and

$$\Pr\left[\frac{\hat{\mu}-\mu}{\sigma} + k_U\frac{\hat{\sigma}}{\sigma} \geq \log\left(-\log\gamma\right)\right] = \alpha\,, \tag{1.111}$$

respectively. Rewriting Eqs. (1.110) and (1.111) as

$$\Pr\left[\frac{\sigma}{\hat{\sigma}}\log\left[-\log\left(1-\gamma\right)\right] - \frac{\hat{\mu}-\mu}{\hat{\sigma}} \geq k_L\right] = \alpha \tag{1.112}$$

and

$$\Pr\left[\frac{\sigma}{\hat{\sigma}}\log\left(-\log\gamma\right) - \frac{\hat{\mu}-\mu}{\hat{\sigma}} \leq k_U\right] = \alpha\,, \tag{1.113}$$

we observe that k_L and k_U are the upper and lower 100% α points of the distributions of the pivotal quantities

$$P_1 = \frac{\sigma}{\hat{\sigma}}\log\left[-\log\left(1-\gamma\right)\right] - \frac{\hat{\mu}-\mu}{\hat{\sigma}}\,,$$

and

$$P_2 = \frac{\sigma}{\hat{\sigma}}\log\left(-\log\gamma\right) - \frac{\hat{\mu}-\mu}{\hat{\sigma}}\,, \tag{1.114}$$

respectively. The distributions of these two pivotal quantities are not derivable explicitly and their percentage points need to be determined either through Monte–Carlo simulations or by approximations.

Mann and Fertig (1973) used the best linear invariant estimators to prepare tables of tolerance factors for Type-II right-censored samples when $n = 3(1)25$ and $n - s = 3(1)n$, where s is the number of largest observations censored in the sample. Thomas *et al.* (1970)

presented tables that can be used to determine tolerance bounds for complete samples up to size $n = 120$, and Billman *et al.* (1972) provide tables which can be used to determine tolerance bounds for samples of sizes $n = 40(20)120$ with 50% or 75% of the largest observations censored. Johns and Lieberman (1966) presented tables that can be used to get tolerance bounds for sample sizes $n = 10, 15, 20, 30, 50$ and 100 with Type-II right censoring at four values of s (the number of observations censored) for each n. Using the efficient simplified linear estimator given in Bain (1972), Mann *et al.* (1974) derived approximate tolerance bounds based on an F-approximation. This F-approximation turns out to be quite effective and can also be utilized with the best linear unbiased estimators μ^* and σ^*; in fact the approximation turns out to be adequate even in the case of moderate sample sizes with heavy censoring.

An alternative F-approximation was proposed by Lawless (1975) for the lower α confidence bound on the quantile X_γ. It is based on the fact that, at least in the case when the censoring in the sample is quite heavy, the estimators $\hat{\hat{\mu}}$ and $\hat{\hat{\sigma}}$ are almost the same as the maximum likelihood estimators $\hat{\mu}$ and $\hat{\sigma}$. This F-approximation is also quite accurate over a wide range of situations. Lawless noted that the quantity

$$Z_\gamma = \frac{1}{\hat{\sigma}}\{\sigma \log(-\log\gamma) - (\hat{\mu} - \mu)\} \tag{1.115}$$

is also a pivotal quantity, since $Z_\gamma = \{\log(-\log\gamma)/Z_2\} - Z_1$ where $Z_1 = (\hat{\mu} - \mu)/\hat{\sigma}$ and $Z_2 = \hat{\sigma}/\sigma$ are pivotal quantities [cf. (1.114)] and can be used to construct tolerance bounds. For example,

$$\Pr[Z_\gamma \geq z_{\gamma,\alpha}] = \alpha \Rightarrow \Pr[z_{\gamma,\alpha}\hat{\sigma} + \hat{\mu} \leq X_\gamma] = \alpha\,, \tag{1.116}$$

and hence $z_{\gamma,\alpha}\hat{\sigma} + \hat{\mu}$ becomes a lower α confidence bound on the quantile X_γ. The percentage points of the distribution of Z_γ in (1.115) therefore yield upper tolerance limits.

Mann and Fertig (1977) discussed the correction for small-sample bias in Hassanein's (1972) asymptotic best linear unbiased estimators of μ and σ based on k optimally selected quantiles. They presented tables of these bias-correction factors for complete samples of sizes $n = 20(1)40$. These tables will allow one to obtain estimates based on the specified sets of order statistics that are best linear unbiased estimates or best linear invariant estimates, and can also be used to determine approximate confidence bounds on X_γ and the related tolerance limits using the approximation approaches mentioned above. Using the conditional method of inference (Sec. 1.7.8), Lawless (1975) has shown that the conditional tail probability of the distribution of Z_γ in (1.115) is given by

$$\Pr[Z_\gamma \geq z|\mathbf{a}] = (n-s-1)!C_{n-s}(\mathbf{a}) \times \int_0^\infty \frac{t^{n-s-2}e^{t\sum_{i=1}^{n-s}a_i}\Gamma_{h(t,z)}(n-s)}{\Gamma(n-s)\{\sum_{i=1}^{n-s}e^{a_i t} + se^{a_{n-s}t}\}^{n-s}}dt\,, \tag{1.117}$$

where $\mathbf{a}$ is the ancillary statistic described in Sec. 1.7.8, $\Gamma_b(p)$ is the incomplete gamma function and

$$h(t,z) = -\log\gamma \cdot e^{-tz}\left\{\sum_{i=1}^{n-s} e^{a_i t} + se^{a_{n-s}t}\right\}. \tag{1.118}$$

The integral in (1.117) is rather complex and needs to be evaluated numerically. The normalizing constant $C_{n-s}(\mathbf{a})$ is determined numerically by using the condition that $\Pr[Z_\gamma \geq -\infty|\,\mathbf{a}] = 1$ (in which case $h(t,z) = \infty$ and $\Gamma_{h(t,z)}(n-s) = \Gamma(n-s)$). Once the percentage points of Z_γ are determined from (1.117) by numerical methods, tolerance limits can be obtained as described above.

Gerisch, Struck and Wilke (1991) used a different approach and discussed the determination of one-sided tolerance limit factors for the exact extreme value distributions from a normal parent distribution. In their opinion, one-sided tolerance limits for the *asymptotic* extreme value distributions cannot be regarded as sufficient approximations of one-sided tolerance limits for the corresponding *exact* extreme value distributions.

A Remark on Prediction

A way of using extreme value theory is, after achieving confidence in a probabilistic model, to use it for *prediction* of the extreme values which are supposed to occur, in the near or far future. This is naturally of a substantial interest, when the problem is to build equipment which has a limited life before failure, and when it is not possible to eliminate completely the possibility of having it being destroyed by some exceptional events.

As Galambos (1981) convincingly demonstrated, this approach is, in most cases, highly unreliable. In fact, very slight variations in the model, accounting for mutual dependence of the random variables or their marginal distributions, may often have dramatic consequences on the prediction of extremes. It seems that prediction of extremes is in general a risky field, in which serious statisticians should be very cautious before taking responsibilities.

1.7.10 *Minimum Distance Estimation of the Gumbel Distribution for Minima*

Consider the cdf

$$G_0(x) = 1 - \exp\left(-\exp\left(x-\mu\right)/\sigma\right); \qquad x \geq \mu, \qquad \sigma > 0. \tag{1.119}$$

As indicated above the distribution $G_0(x)$ plays a central role as a limiting distribution of the minima m_n of a sequence of i.i.d. random variables Y_i, $i \leq n$, as $n \to \infty$. Let $F_n(\cdot)$ be the empirical cdf. The minimum distance estimators of (μ, σ) minimize the unweighted Cramér–von–Mises distance

$$(\mu^*, \sigma^*) = \arg\min_{\mu,\sigma} \int_{-\infty}^{\infty} \left(F_n(x) - G_0\left(\frac{x-\mu}{\sigma}\right)\right)^2 dx\,.$$

If the minimum of the Cramér–von–Mises distance exists, μ^* and σ^* are called the minimum distance (MD) estimators of the location and scale parameters respectively. They are solutions of the equations

$$\sum_{i\leq n}\left(G_0\left(\frac{X_i-\mu}{\sigma}\right) - \frac{1}{2}\right) = 0$$

and

$$\sum_{i\leq n}\left(k_0\left(\frac{X_i-\mu}{\sigma}\right) - c_k\right) = 0\,,$$

with $k_0(x) = \int_{-\infty}^{x} y g_0(y) dy$, where g_0 is the density of G_0 and constant $c_k = -(\gamma + \ln 2)/2$. (Here γ denotes Euler's constant.) Moreover, $\sqrt{n}((\mu^*, \sigma^*) - (\mu, \sigma)) \to N(0, \Sigma^*)$, where the covariance matrix Σ^* is given by

$$\Sigma^* = \sigma^2 \begin{pmatrix} 1.1801 & -0.1758 \\ -0.1758 & 0.7953 \end{pmatrix} \qquad \text{(see, e.g., Dietrich and Husler (1996)).}$$

Dietrich and Husler (1996) have also shown that

$$\text{ARE}\,(\mu^*, \hat{\mu}) = 0.9395\,,$$
$$\text{ARE}\,(\sigma^*, \hat{\sigma}) = 0.7644\,,$$

where $(\hat{\mu}, \hat{\sigma})$ are maximum likelihood (ML) estimators. Thus the MD location estimator μ^* is quite efficient but the MD scale estimator σ^* is less so. However, MD estimators are robust and have bounded influence function; consequently the very extreme values have less influence on the MD estimators as compared with the ML ones. In fact, Dietrich and Husler (1996) have shown that the *breakdown point* of the MD estimator of the location parameter is 0.5 and of the scale parameter is 0.2026. Thus it is expedient to use MD estimators if there is suspicion that the data may be contaminated. (Compare with the section on *Robust Estimation* in Chap. 2.)

1.8 Distributions Related to the Classical Extremal Distributions

There is clearly a close connection between the three types of extremal distributions. The standard type 1 extreme value distribution is a transitional limiting form between type 2 (Fréchet) and type 3 (Weibull) distributions. Furthermore, a logarithmic transformation of a Weibull random variable results in a type 1 extreme value random variable. Also, as noted earlier, if Y is a standard type 1 extreme value random variable with density (1.22), then e^{-Y} has a standard exponential distribution.

A rather unexpected relation holds between the logistic and type 1 distributions. If two independent random variables each have the same type 1 distribution, their difference has a logistic distribution given by $F(z) = 1 - [1 + \exp((x-\mu)/\sigma)]^{-1}$ with $\sigma > 0$. [Gumbel (1961)]. Gumbel (1962c, d) has also studied the distribution of products and ratios of independent variables having extreme value distributions. We shall return to this topic in the sequel.

1.8.1 *Limiting Distributions of the rth Greatest (Least) Value*

Limiting distributions of second, third, and so forth, greatest (or least) values may be regarded as being related to extreme value distributions. Gumbel (1958) has shown that under the same conditions as those leading to the type 1 extreme value distribution, the limiting distribution of the rth greatest value $Y'_{n-r+1} = (X'_{n-r+1} - \mu)/\sigma$ has the standard form of probability density function

$$p_{Y'_{n-r+1}}(y) = r^r[(r-1)!]^{-1} \exp[-ry - re^{-y}]\,. \qquad (1.120)$$

$100\alpha\%$ points of this distribution are given by Gumbel (1958) to five decimal places for

$$r = 1(1)15(5)50\,,$$
$$\alpha = 0.005, 0.01, 0.025, 0.05, 0.1, 0.25, 0.5, 0.75, 0.9, 0.95, 0.975, 0.99, 0.995\,.$$

The moment-generating function of distribution (1.120) is

$$\frac{r^t\Gamma(r-t)}{\Gamma(r)}\,.$$

The cumulant-generating function is

$$t\log r + \log\Gamma(r-t) - \log\Gamma(r)\,,$$

and the cumulants are

$$\kappa_1 = \log r - \psi(r)\,, \tag{1.121}$$
$$\kappa_s = (-1)^r\psi^{(s-1)}(r)\,, \qquad s \ge 2\,.$$

It is important to note that the limiting distribution (1.120), which corresponds to a fixed value of r, should be distinguished from distributions obtained by allowing r to vary with n (usually in such a way that r/n is nearly constant) or keeping r constant but varying the argument value. Borgman (1961), for example, has shown that if x_n be defined by $F_X(x_n) = 1 - w/n$, for given fixed w [where $F_X(x)$ is the cdf of the population distribution], then

$$\lim_{n\to\infty} \Pr[X'_{n-r+1} \le x_n] = 1 - [(r-1)!]^{-1}\int_0^w t^{r-1}e^{-t}dt\,. \tag{1.122}$$

Note that the right-hand side of (1.122) can also be written in terms of a χ^2 distribution, as $\Pr[\chi^2_{2r} > 2w]$.

1.8.2 *The Asymptotic Distribution of Range*

The asymptotic distribution of *range* is naturally closely connected with extreme value distributions. If both the greatest and least values have limiting distributions of type 1, then [Gumbel (1947)] the limiting distribution of the range, R, is of the form

$$\Pr[R \le r] = 2e^{-r/2}K_1(2e^{-r/2})\,, \qquad r > 0\,, \tag{1.123}$$

with probability density function

$$p_R(r) = 2e^{-r}K_0(2e^{-r/2})\,, \qquad r > 0\,,$$

where K_0, K_1 are modified Bessel functions of the second kind of orders zero, one, respectively. Explicitly:

$$K_0(z) = -\left\{\ln\left(\frac{1}{2}z\right) + \gamma\right\} I_0(z) + \frac{\frac{1}{4}z^2}{(1!)^2}$$
$$+ \left(1 + \frac{1}{2}\right)\frac{(\frac{1}{4}z^2)^2}{(2!)^2} + \left(1 + \frac{1}{2} + \frac{1}{3}\right)\frac{(\frac{1}{4}z^2)^3}{(3!)^2} + \cdots\,,$$

where

$$I_0(z) = 1 + \frac{\frac{1}{4}z^2}{(1!)^2} + \frac{(\frac{1}{4}z^2)^2}{(2!)^2} + \frac{(\frac{1}{4}z^2)^3}{(3!)^2} + \cdots$$

and

$$K_1(z) = (\gamma - \log 2 + \log z) \sum_0^\infty \frac{1}{v!(v+1)!} \left(\frac{z}{2}\right)^{2v+1}$$
$$+ \frac{1}{z} - \sum_1^\infty \frac{1}{(v-1)!v!} \left(\frac{z}{2}\right)^{2v-1} \left(S_v - \frac{1}{2v}\right),$$

where

$$S_v = \sum_{\lambda=1}^{v} 1/\lambda .$$

Note that $K_1(z) = -K_0'(z)$.

Gumbel (1947) gave the values

$$E[R] = 2\gamma = 1.15443 ,$$
$$\text{median } R = 0.92860 ,$$
$$\text{modal } R = 0.50637 .$$

Also

$$\text{var}(R) = \frac{\pi^2}{3} = 3.2899 .$$

In Gumbel (1949b), there are tables of $\Pr[R \le r]$ and $p_R(r)$ to seven decimal places for

$$r = -4.6(0.1) - 3.3(0.05)11.00(0.5)20.0 ,$$

and of percentile points R_α to four decimal places for

$$\alpha = 0.0002(0.0001)0.0010(0.001)0.010(0.01)0.95(0.001)0.998$$

and to three decimal places for

$$\alpha = 0.0001, 0.999(0.0001)0.9999 .$$

Further details on the asymptotic distribution of range are given in Gumbel's book (1958) and in Galambos (1987).

1.8.3 *Extremal Quotient*

Let $M_n = \max\{X_1, \ldots, X_n\}$ and $m_n = \min\{X_1, \ldots, X_n\}$ where $\{X_n : n \ge 1\}$ is a sequence of i.i.d. random variables.

The extremal quotient is defined by

$$Q = \frac{M_n}{m_n}$$

(see Gumbel and Herbach (1951)). Gumbel (1958) defined this quotient M_n/m_n under the assumption that $m_n < 0$. One of the earlier uses is in climatology (Carnard (1946)). The quotient is scale-invariant. Gumbel and Herbach (1951) derived the exact form of the cdf of this statistic. The cdf $H_\lambda(q)$ of the extremal quotient is

$$H_\lambda(q) = [\lambda/(1-e^{-\lambda})^2] \int_0^1 \exp[-\lambda(z+z^q)]dz - e^{-\lambda}/(1-e^{-\lambda}),$$

where

$$\Pr[M_n \le x] = G_n(x) \qquad \text{converges to } \exp(-e^{(x-\mu_n)/\sigma_n}) \qquad (\text{for large } n),$$

and $\lambda = \exp(\mu_n/\sigma_n)$ is a function of the initial distribution and the size n of the sample from which the quotient was drawn. The parameter λ is dimensionless. For large λ the distribution function becomes

$$H_\lambda(q) = \lambda \int_0^1 \exp(-\lambda(z+z^q))dz.$$

The distribution of the extremal quotient rapidly becomes concentrated with increasing sample size. The concentration is about the median, which is unity. In order to compensate for this concentration, the difference $Q-1$ is multiplied by μ_n and σ_n. Thus we have the variable

$$\tau \equiv (Q-1)\log\lambda.$$

It was shown by Gumbel and Keeny (1950) that the distribution of τ approaches the logistic distribution as λ tends to infinity; i.e. for all x,

$$\lim_{\lambda\to\infty} P\{\tau \le x\} = 1/(1+e^{-x}).$$

It should be noted, however, that while a logistic variate has all the moments, τ has none (since Q has none). Gumbel and Pickand (1967) traced the extremal quotient on logarithmic normal paper for $\lambda = 2$, 5, 10, 20, 100 and 492.7 (see Graph 1).

The curve for $\lambda = 1,000$ is indistinguishable from that for $\lambda = 492.7$. The distribution function of the extremal quotient plots nearly as a straight line for large values of λ, although the moments of the extremal quotient do not exist.

The asymptotic distribution function when the initial distribution function is of exponential or Cauchy types were studied by Gumbel and Keeny (1950). Tables of the distribution of "extremal quotient" were published by Gumbel and Pickands (1967).

Recently, Bakarat (1998) obtained necessary and sufficient conditions for the weak convergence of sample extremal quotient of i.i.d. random variables as $n \to \infty$. For type 1 (Gumbel) and type 2 (Fréchet) limit distributions of M_n and m_n the extremal quotient Q, properly normalized, converges weakly to distribution function

$$Q(q;\alpha,\beta) \cdot I_{(-\infty,0)}(q) + I_{[0,\infty)}(q)$$

where $I_A(\cdot)$ is the indicator function of the set A and

$$Q(q;\alpha,\beta) = 1 - \int_0^\infty \exp(-y - |q|^{-\beta} y^{\beta/\alpha})dy,$$

Graph 1
Distribution function of the extremal quotient

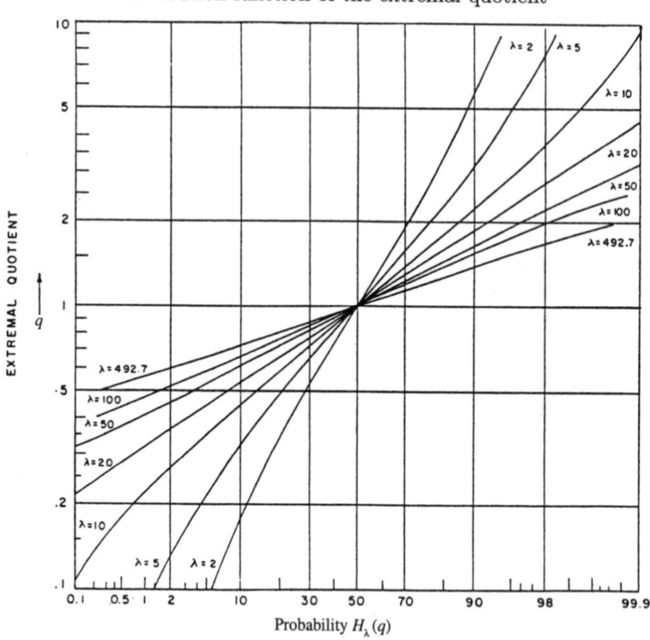

where α and β are positive constants appearing in the definitions of type 1 distributions: $(\exp -y^{-\beta})$ $(y \geq 0)$; $1 - \exp(-(-x)^{-\alpha})$ $(x < 0)$ and type 2 distributions: $\exp(-(-y)^{\beta})$ $(y < 0)$; $1 - \exp((-x)^{\alpha})$ $(x \geq 0)$.

Gumbel and Keeney (1950) proposed to estimate the parameter λ by comparing the expected proportion of the sample for which $1/2 < Q < 2$, with the observed proportion. However, unless λ is extremely small, all of the sample will lie in this range with very high probability. So the method is not always applicable.

The cumbersome nature of the distribution function makes it plain that even on a modern computer, maximum likelihood estimation would not be easy. But since Q has no moments, and the median is 1, independent of λ, neither quantile nor moment methods will be appropriate. A further study of this basic statistic is desirable.

1.8.4 *Log-Gamma Density*

The standard log-gamma density function

$$p_Y(y) = \frac{1}{\Gamma(\kappa)} e^{\kappa y - e^y}, \qquad -\infty < y < \infty, \qquad \kappa > 0 \tag{1.124}$$

can be viewed as a generalization of the standard type 1 extreme value density.

Specifically, if Y has the density function in (1.124), for the case when $\kappa = 1$ the variable $-Y$ is distributed as a standard type 1 extreme value random variable. We note that for integral values of κ, density (1.124) is related to the density (1.120). The cumulative distribution function corresponding to the density (1.124) is

$$F_Y(y) = I_{e^y}(\kappa)\,, \qquad -\infty < y < \infty\,, \qquad \kappa > 0\,, \tag{1.125}$$

where $I_t(\kappa)$ is the incomplete gamma function ratio

$$I_t(\kappa) = \int_0^t \frac{1}{\Gamma(\kappa)} e^{-z} z^{\kappa-1}\, dz\,, \qquad 0 < t < \infty\,, \kappa > 0\,.$$

For integral values of κ, therefore, we have

$$1 - F_Y(y) = e^{-e^y} \sum_{i=0}^{\kappa-1} \frac{e^{iy}}{i!}\,, \quad \infty < y < \infty\,, \quad \kappa = 1, 2, \ldots \tag{1.126}$$

(this well-known relation can easily be verified by successive differentiation). The moment-generating function corresponding to the density (1.124) is

$$E[e^{tY}] = \Gamma(\kappa + t)/\Gamma(\kappa)\,;$$

in particular, we have

$$E[Y] = \psi(\kappa) \qquad \text{and} \qquad \text{var}(Y) = \psi'(\kappa)\,. \tag{1.127}$$

Since the digamma function $\psi(\kappa) \sim \log \kappa$ and $\psi'(\kappa) \sim 1/\kappa$ for large κ, Prentice (1974) suggested a reparametrized log-gamma density function

$$p_Y^*(y) = \frac{\kappa^{\kappa-1/2}}{\Gamma(\kappa)} e^{\sqrt{\kappa}y - \kappa e^{y/\sqrt{\kappa}}}\,, \qquad -\infty < y < \infty\,, \qquad \kappa > 0 \tag{1.128}$$

which tends to the standard normal density function as $\kappa \to \infty$. By introducing a locating parameter μ and a scale parameter σ in the density (1.124) we obtain a three-parameter log-gamma density function as

$$p_X(x) = \frac{1}{\sigma\Gamma(\kappa)} e^{\kappa(x-\mu)/\sigma} e^{-e^{(x-\mu)/\sigma}}\,, \quad \kappa > 0\,, \quad \sigma > 0\,. \tag{1.129}$$

This is evidently a generalization of the type 1 extreme value density function (1.21). Lawless (1980, 1982) has illustrated the usefulness of the three-parameter log-gamma density (1.129) as a life-test model and discussed maximum likelihood estimation of the parameters. Balakrishnan and Chan (1994a, b, c, d) have studied order statistics from this distribution and also the BLU and, the asymptotic BLU estimations, as well as the maximum likelihood estimation of the parameters based on complete and Type-II censored samples. Young and Bakir (1987) have discussed the log-gamma regression model.

1.8.5 *Smallest Extreme Value (SEV) Regression*

SEV regression model has received special attention in reliability applications in particular in accelerated life testing. This model uses the smallest extreme value distribution to describe the variability in product's (log)lifetime at a particular stress level and assumes a linear relationship between (log)life and the transformed stress variable.

Specifically the model states that for a given value x of independent (response) variable, the r.v. Y follows a SEV distribution with parameter $\mu(x) = \gamma_0 + \gamma_1 x$ and scale parameter $\sigma > 0$. (Evidently γ_0 and γ_1 are respectively the intercept and slope parameters.)

Hence,

$$f(y|x) = \frac{1}{\sigma} \exp\left[\frac{y - (\gamma_0 + \gamma_1 x)}{\sigma} - \exp\left(\frac{y - (\gamma_0 + \gamma_1 x)}{\sigma}\right)\right]$$

with

$$-\infty < y < \infty, \qquad -\infty < \gamma_0 < \infty, \qquad -\infty < \gamma_1 < \infty \quad \text{and} \quad \sigma > 0.$$

Here, γ_0, γ_1 and σ are estimated from sample data which may include complete (as well as censored) observations of Y. The q quantile at a given value of x, say x_D, is

$$\begin{aligned} y_q(x_D) &= \mu(x_D) + \{\log[-\log(1-q)]\}\sigma, \\ &= \gamma_0 + \gamma_1 x_D + \{\log[-\log(1-q)]\}\sigma. \end{aligned}$$

It is known that the location parameter $\mu(x_D)$ is also the 0.632 quantile of the SEV distribution at x_D (the so-called nominal (log)life of the product at x_D).

As mentioned above, in accelerated life-testing applications, the SEV distribution is used to describe the scatter in the product's (log) lifetimes at a particular stress level. Furthermore, many accelerated live models which are of the form $\mu(x) = \gamma_0 + \gamma_1 x$ express the nominal (log)life $\mu(x)$ as a linear function of a (possibly transformed) stress variable x. For example, the *inverse power law* states that $x = \log(V)$ where V is the voltage. In the *Arrhenius relationship* $x = 1/T$ where T is the absolute Kelvin temperature. The scale parameter σ is assumed constant for all x.

Doganaksoy and Schmee (1991) constructed and compared various approximations to confidence intervals for the SEV distribution simple linear regression model under time censoring. Intervals based on the asymptotic normality of MLE are "anti-conservative" and caution is needed in using them. On the other hand, uncorrected likelihood ratio intervals are remarkably accurate in situations with heavy censoring.

Distributions related to the generalized extreme value distributions are discussed at the end of Sec. 2.

1.9 Applications of the Classical Extreme Value Distributions

The range of applications of extreme value distributions is extremely (no pun intended) wide, and it is a daunting task to list all of them without subjecting our readers to a boring experience. We appeal to their patience, curiosity and perseverance to carefully review the next few pages. To highlight the applications we just indicate such diverse areas as break frequency of paper, horse racing, network design, queues in supermarkets, synthetic membranes, sizes of bush fires, not to mention the obvious topics such as high temperatures, earthquakes, risk management, winds, floods, ozone concentration, insurance and more recently, financial matters. The list below, being a substantially updated (and in a sense condensed) version of Section 14 of the Chapter on Extreme Value Distributions in Johnson, Kotz and Balakrishnan's book on *Continuous Univariate Distribution* (J. Wiley, 1995) attempts to provide a meaningful and hopefully coherent picture. Additional applications are mentioned in the section on generalized extreme value distributions and in the second part of the monograph dealing with the multivariate extreme value distributions. Unavoidable coloring of choice by personal taste may have done injustice – omitting worthwhile contributions.

As mentioned earlier, E. J. Gumbel played a pioneering role during the 40s and 50s in bringing out several interesting applications for the extreme value data and developing sound statistical methodology. We shall briefly describe below some outstanding applied papers in a more or less chronological order.

Probably the first paper that described an application of extreme values in flood flows was by Fuller (1914). Griffith (1920) brought out an application while discussing the phenomena of rupture and flow in solids. Next, Gumbel (1937a, b) used the extreme value distribution to model radioactive emissions and human lifetimes. The use of the distribution to model the rupture in solids was discussed by Weibull (1939). Weibull effectively advocated the use of reversed type 3 distributions which have now become widely known as *Weibull distributions.*

Gumbel (1941) applied the distribution to analyzing data on flood flows, and in subsequent works he continued his discussion on the plotting of flood discharges, estimation of flood levels, and forecast of floods [Gumbel (1944, 1945, 1949a)]. The application to study earthquake magnitudes was pointed out by Nordquist (1945). Velz (1947) used the distribution to model microorganism survival times. Epstein (1948) applied the theory of extreme values to problems involving fracture data. Rantz and Riggs (1949) illustrated an application while analyzing the magnitude and frequency of floods in the Columbia River Basin measured in the course of a U.S. Geological Survey. An interesting new application of the extreme value distribution was used by Potter (1949) to study rainfall data. Weibull (1949) emphasized the role of extreme value distributions to represent fatigue failures in solids and advocated once again the use of the Weibull distribution in place of the type 1 extreme value distribution.

In meteorology, the popularity of the type 1 distribution is due mainly for the following six reasons:

(1) The EV1 distribution results from an initial (unlimited) distribution of exponential type which converges to an exponential function;

(2) under certain assumptions, the extreme values in a sample follow this distribution;

(3) it is simple and has only two parameters;

(4) from a statistical viewpoint, it may be preferable to apply even when the sample size is small;

(5) because it is available in closed form, it is easier to determine the extreme value for a specified value of probability or return period; and

(6) in a Monte Carlo study, the EV1 random variables can be easily generated.

The so-called Gumbel method has been applied successfully to both regular-type events (e.g., temperature and vapor pressure) and irregular-type events (e.g., rainfall and wind). Thom (1954) emphasized that the sparse sampling in time of extreme events obscures much of the information in a rainfall process. Methods of analysis of extreme hydrological events have changed gradually since the publication of Gumbel (1941) on asymptotic theory dealing with flood discharges by streams. Until quite recently basic assumptions of the theory were that the frequency distribution of extremes within successive intervals remains constant and that observed extremes may be viewed as being independent samples from a homogeneous population.

Gumbel (1954, 1958) presented consolidated accounts of the statistical theory of extreme values and several practical applications. These works may be studied in conjunction with his later works [Gumbel (1962a, b)] to gain a deeper understanding and better knowledge of extreme value distributions.

Longuet-Higgins (1952) contribution is one of the earliest works on heights of sea waves in the framework of extreme value analysis. Thom (1954) (mentioned above) applied the distribution while discussing the frequency of maximum wind speeds. Numerous papers by Thom scattered in diverse publications on wind and waves applications appeared in the late '60s and early '70s. Aziz (1955, 1956) applied the extreme value theory to an analysis of maximum pit depth data for aluminum. Kimball (1955) — mentioned above — explained several practical applications of the theory of extreme values and also described some aspects of the statistical problems associated with them. Jenkinson (1955) followed Potter (1949) by applying the extreme value distribution to model the annual maximum or minimum values of some meteorological elements. Lieblein and Zelen (1956) carried out an extensive study related to inference based on the extreme value distribution and applied their methods to investigate the fatigue life of deep-grove ball bearings. Eldredge (1957) discussed an analysis of corrosion pitting by extreme value statistics and applied it to oil well tubing caliper surveys. King (1959) summarized developments on extreme value theory and explained their implications to reliability analysis. Metcalfe and Smith (1964) investigated applications to glass fibers. Clough and Kotz (1965) presented some queuing model applications for the extreme value distributions. Posner (1965) — mentioned above — detailed an application of the extreme value theory to communication engineering; see also the comments by Gumbel and Mustafi (1966) on this paper. In a series of reports Simiu and Filliben (1975, 1976) and Simiu *et al.* (1978) used extensively the extreme value distributions in the statistical analysis of extreme winds. Regional flood frequency analysis based on the type 1 (Gumbel) distribution using Bayesian estimation was carried in 1971 by Cunnane and Nash (1974).

Shen *et al.* (1980) applied the distributions for predictions of flood. Watabe and Kitagawa (1980) demonstrated an application while discussing the expectancy of maximum earthquake motions in Japan. Okubo and Narita (1980) used the extreme value distribution to model the data on extreme winds in Japan. Wantz and Sinclair (1981) carried out a similar analysis on the distribution of extreme winds in the Bonneville power service area. Metcalfe and Mawdsley (1981) applied extreme value distribution to estimate extreme low flows for pumped storage reservoir designs. The use of the distribution in regional flood frequency estimation and network design was illustrated by Greis and Wood (1981). Roldan-Canas, Garcia-Guzman, and Losada-Villasante (1982) constructed a stochastic extreme value model for wind occurrence. A comprehensive application of the extreme value distribution in rainfall analysis was provided by Rasheed *et al.* (1983). Henery (1984) presented an intriguing application of the extreme value model in predicting the results of horse races. While Pericchi and Rodriguez-Iturbe (1985) used the extreme value distribution in a statistical analysis of floods, Burton and Makropoulos (1985) applied it in an analysis of seismic risk of circum-Pacific earthquakes. The usefulness of this distribution to model time-to-failure data in reliability studies has been discussed by Canfield (1975) and Canfield and Borgman (1975).

A two-component extreme value distribution was proposed by Rossi *et al.* (1986) for flood frequency analysis; also see the comments on this paper by Beran *et al.* (1986) and Rossi's (1986) subsequent reply. J. A. Smith (1987), Jain and Singh (1987), and Ahmad, *et al.* (1988) provided further discussions on the application of the type 1 extreme value distribution for flood frequency analysis. Achcar *et al.* (1987) discussed the advantages of transforming a survival data to a type 1 extreme value distribution form before analyzing it. Nissan (1988) demonstrated an early application of the type 1 distribution in estimating insurance premiums. The role of statistics of extremes in climatological problems was discussed by Buishand (1989). A seminal paper by Smith (1989) is devoted to trend detection in ground level ozone.

Cockrum *et al.* (1990) and Taylor (1991) applied the extreme value distributions in modelling and simulation involving product flammability testing. Wiggins (1991) displayed an earlier application in stock markets. A mixture of extreme value distributions was used by Fahmi and Abbasi (1991) to study earthquake magnitudes in Iraq and conterminous regions. Tawn (1992) discussed the estimation of probabilities of extreme sea levels, while Hall (1992) discussed further on flood frequency analysis. Tawn's numerous pioneering applications are discussed in the sections on generalized extreme value distribution and multivariate extreme value distributions. Bai *et al.* (1992) demonstrated an application of the extreme value distribution in predicting the upper percentiles that are of interest in environmental quality data.

Hopke and Paatero (1993) discussed the extreme value estimation in the study of airborne particles. Kanda (1993) considered an empirical extreme value distribution to model maximum load intensities of the earthquake ground motion, the wind speed, and the live load in supermarkets. Goka (1993) applied the extreme value distribution to model accelerated life-test data to tantalum capacitors for space use and to on-orbit data of single event phenomenon of memory integrated circuits in the space radiation environment. Rajan (1993) stressed on the importance of the extreme value theory by providing experimental examples where significant deviations from the average microstructure exist in pertinent materials physics (in particular pore size distributions in synthetic membranes). Scarf and Laycock (1993) and Shibata (1993) have demonstrated applications of extreme value theory in corrosion engineering. Applications of extreme values in insurance have been illustrated by Teugels and Beirlant (1993) in their pioneering paper and a subsequent monograph co-authored with Vynckier (1996). Diebold *et al.* (1999) provide a balanced assessment of the use of extreme value theory in risk management.

Dasgupta and Bhaümik (1995) discussed lethal effects of the ozone depletion and computed the upper and lower β-content confidence limits for an extreme value distribution showing that these can be used to calculate the upper and lower tolerance limits to the level of atmospheric ozone layer. Their methodology was described earlier in the section on tolerance limits. They use the data of Pallister and Tuck (1983) as presented in Pyle (1985) consisting of percentage deviation from midnight values of ozone concentration for a diurnel cycle.

Sizes of bush fires observed in Australia in 1986–1987 reported by the Environmental Protection Agency (EPA) were analyzed by Smith (1993). Both Fréchet and Gumbel type distributions were fitted. Only fires that burned an area of 1 hectare or more were recorded. Thus 75 recorded fires should be viewed as the largest observations from a sample of a large size.

The data for annual maximal winds for Jacksonville, Florida (stored in the file em-jwind.dat) was recorded by Changery (1982) (and discussed in Kinnison (1985)) for the years 1950–1979. The range of the data is maximal wind speed between 34–74 mph (in the years 1959 and 1964 respectively) with a pronounced mode of 42 mph. The data was subdivided into tropical and non-tropical storm years with seven observations in the latter. The Gumbel model yielded MLE estimators

$$(\mu, \sigma) = (43.6, 6.7) \quad \text{for the tropical}$$

and

$$(\mu, \sigma) = (44.1, 9.0) \quad \text{non-tropical storm data}\,.$$

Viewing the tropical maximum annual wind speeds as randomly left-censored by non-tropical ones, Reiss and Thomas (1997) observed that the distribution is now shifted to the left and tropical wind speeds are now better described by a Fréchet density indicating a heavier upper tail, which may mislead forecast of catastrophic tropical storms.

Some most recent applications as of this writing include:

(1) Transforming point rainfall into areal rainfall to obtain relationships known in meteorology as intensity-duration frequency curves (Sivapalan and Bloschl (1998)) (type 1 distribution),

(2) extreme occurrences in Germany's stock index (Broussard and Booth (1998)),

(3) behavior of solar proton peak fluxers (Xapsos *et al.* (1998)) (type 2 distribution),

(4) probabilities of *grant* freak waves in areas surrounding Japan's seacoast (Yasuda and Mori (1997)),

(5) discussion of pitfalls and opportunities in the use of extreme value theory in risk management (Diebold *et al.* (1999)).

A more detailed list of the most recent application is provided by Nadarajah (2000). Proceedings of the Gaithersburg (MD, U.S.A.) Conference edited by J. Galambos *et al.* (1994) and the Conference on Stochastic and Statistical Methods in Hydrology and Environmental Engineering edited by K. W. Hisel (1994) constitute a most valuable collection of investigations devoted to applied aspects of extreme value analysis. The book by Embrechts *et al.* (1997) is an excellent source for theory and applications in insurance and finance — the currently most glamorous fields of extreme value analysis. Data examples provided in Castillo (1988) are most valuable for applications.

Appendix to Chapter 1

A. Some Comments on Gnedenko's Results

For the readers — experts in probability theory — we note that Gnedenko's condition for the type 2 extreme value distribution discussed in section 1 is equivalent to the condition that the sum $S_n = X_1 + \cdots + X_n$ belongs to the domain of attraction of a stable law with characteristic exponent a, where $0 < a < 2$, with the characteristic function given by:

$$\phi(u) = \exp\left(i\gamma u - c|u|^a\{1 + i\beta(t/|t|)\omega(u,a)\}\right), \qquad \text{where}$$

$$\omega(u,a) = \tan(\pi a/2) \quad \text{for } a \neq 1, \qquad \omega(u,a) = (2/\pi)\log(u) \quad \text{for } a = 1.$$

Here γ, c and β are appropriate constants.

(Recall that a *necessary* condition for type 1 limiting distribution is that, for any $c > 0$,

$$\lim_{x\to\infty}\left\{\left(\frac{1-F(x)}{1-F(cx)}\right)\right\} = 0 \qquad \text{provided } x_0 \equiv \max\{t; F(t) < 1\} = \infty.)$$

The Poisson distribution satisfies neither $\lim_{x\to\infty}\{(1-F(x))/(1-F(cx))\} = 0$, $c > 0$, nor the condition $\lim_{x\to\infty}\{(1-F(x))/(1-F(cx))\} = c^k$, i.e. it does not belong to the domain of attraction of an extreme value distribution. The same is true for some other discrete distributions. See Anderson (1970) who discusses conditions for a class of discrete distributions.

For a normally distributed sequence, $X_1, X_2, \ldots, X_n$ of i.i.d. random variables $N(0,1)$ with $Y_n = \max\{X_i\}$, $i = 1, \ldots, n$, we have (see, e.g., Cramér (1946), p. 475).

$$\lim_{n\to\infty} P[\sqrt{2\log n}\{Y_n - \sqrt{2\log n} + ((\log\log n + \log 4\pi)/2\sqrt{2\log n})\} < x] = e^{-e^{-x}}.$$

The result implies that

$$\lim_{n\to\infty}\{Y_n - \sqrt{2\log n}\} = 0 \quad \text{in probability.}$$

(This is an example of the so-called *stability* of $\{Y_n\}$ sequence.)

The interrelation between the respective domains of attraction of Y_n, $Z_n = \min(X_1, \ldots, X_n)$ and S_n was investigated in detail by Rosengard (1962) and Tiago de Oliveira (1962), among others. In the case when $\mathrm{Var}(X_i) < \infty$, Y_n, Z_n and S_n are asymptotically independent.

For many well-known distributions (including the normal), the limiting distribution of $P\{(Y_n - b_n)/a_n \leq x\} = F^n(a_n x + b_n)$, is of type 1. (Here, as usual, $Y_n = \max_i X_i$, where $X_1, \ldots, X_n$ are independent random variables with a common c.d.f. F.) This property is sometimes used as a theoretical justification for the adoption of type 1 rather than types 2 or 3. However, for the case of normal extremes, in which $F = \Phi$, the distribution function of a standard normal variable, Fisher and Tippett (1928) showed empirically that the type 3 approximation is closer to $\Phi^n(x)$ than the (limiting) type 1 approximation.

It should also emphasized that *a priori* there is no reason to believe that empirical distributions ought to have tails such that the distribution of normed maxima should converge to some stable type. In fact, maxima can have any distribution and, for the same underlying distribution, the distribution of maxima for a certain sample size may be completely different from that for some other sizes.

In fact, Green (1976) has showed that tails of distributions do not have to be such that the maxima of the random variables they govern will approach some stable limiting distribution. (This "anti-extreme-value" sentiment has not, however, deterred applied researchers from applying the theory described in this book to a multitude of types of empirical data.)

Fisher and Tippett (1928) called the approximation of the form

$$P_n(x) = \begin{cases} \exp\{-(-Ax+B)^k\} & Ax < B\,, \\ 0 & Ax \geq B\,, \quad A < 0 \\ 1 & Ax \geq B\,, \quad A > 0\,, \end{cases}$$

where A, B and k are sequences of parameters depending only on n and $Ak > 0$, the *penultimate* approximation.

Cohen (1982a) provided bounds on $\sup_x |\Phi^n(x) - P_n(x)|$ for special choices of k, A and B and their relation to a_n and b_n and shows that the sup is $O(a_n^4)$. He also proves rigorously that the penultimate approximations provides substantially better approximations than the type 1 approximation $\Lambda\{(x-b_n)/a_n\}$ where $\Lambda(x) = \exp(-e^{-x})$, even for small n. Cohen's (1982) proof is very delicate and involves several refined inequalities.

Hall (1980) in an equally important paper shows essentially that approximations to $\Phi^n(x)$ based on inequalities for the normal tail function are much closer than the penultimate approximation. Thus if the X_i's are indeed independent and identically normally distributed and if n is known, then Hall (1980) provides better estimates of the distribution of $Y_n = \max_i\{X_i\}$ than the approximations based on extreme value theory. However, in practice we are very often uncertain of the normality, the independence and also the value of n. Since the three limit laws apply to a large class of initial distributions, and quite often in certain dependent cases, extreme value theory approximations are more robust than the alternatives suggested by Hall (1980).

In a subsequent paper [Cohen (1982b)] the author extended the above result and reached the somewhat controversial conclusion that there are very good theoretical reasons in certain statistical situations for fitting type 2 and type 3 extreme-value distributions to the observed extremes, even if it is suspected that the limiting form is type 1, *unless* the amount of data available is small. Similar, independent results were obtained by Gomes (1984).

The speed of convergence of $F^n(a_nx+b_n)$ (with optimal normalizing constants a_n, b_n) towards the type 1 extreme value distribution $\Lambda(x)$ has been evaluated by Hall (1979). He showed that there exist contains C_1, C_2 (independent of n) such that

$$\frac{C_1}{\log n} < \sup_x |F^n(a_nx+b_n) - \Lambda(x)| < \frac{C_2}{\log n}\,.$$

This result shows that in the normal case, convergence to Gumbel's Λ type 1 extreme value distribution is rather slow. This phenomenon occurs frequently enough to become a drawback to a careless use of extreme value distributions when it is known that they are generated by small samples.

Cheng *et al.* (1998) investigated almost sure convergence in extreme-value theory. Let $G(\cdot)$ be one of the extreme-value distributions and as usual $Y_n = \max X_i$, where X_i ($i = 1, 2, \ldots, n$) are independent random variables with a common cdf F.

Assume $F \in D(G)$, i.e. there exist $a_n > 0$ and $b_n \in R$ such that

$$P\{(Y_n - b_n)/a_n) \leq x\} \to G(x)\,, \qquad \text{for } x \in R\,.$$

Let $1_{(-\infty,x]}(\cdot)$ denote the indicator function of the set $(-\infty, x]$ and $S(G) =: \{x : 0 < G(x) < 1\}$, the support of G. Obviously $1_{(-\infty,x]}\,((Y_n - b_n)/a_n)$ does not converge almost surely for any $x \in S(G)$.

The same authors also proved that:

$$P\left\{\lim_{N\to\infty}\sup_{x\in S(G)}\left|\frac{1}{\log N}\sum_{n=1}^{N}\frac{1}{n}1_{(-\infty,x]}((Y_n-b_n)/a_n)-G(x)\right|=0\right\}=1\,.$$

Barakat (1997) discussed continuation of weak convergence of suitably normalized extremes from a finite interval $[a,b]$ to the whole real line. The weak convergence of the extremes to a limiting type is, in some sense, stronger than what could be expected. It was shown by Pickands (1968), that if $(Y_n-b_n)/a_n$ did converge to a limiting extreme value distribution, then the moments of $(Y_n-b_n)/a_n$, provided they exist, converge to the corresponding moments of the limiting distribution.

Lucenó (1994) investigated the speed of convergence of the distribution of normalized maximum of a sample of i.i.d. random variables to its asymptotic distribution measuring the difference on the double log-scale graph paper. The convergence to the asymptotic distribution may not be uniform on this scale and the difference between the actual and asymptotic distributions, on the probability plotting paper, may be a logarithmic, power, or even an exponential function in the upper tail when the latter distribution is of Gumbel type 1, but that difference is at most logarithmic in the upper tail for type 2 and 3 distributions.

Gnedenko's (1943) results were generalized by Smirnov (1949). For every k $(1\le k\le n)$ denote by X_{kn} the r.v. that assumes the kth value in descending order of magnitude among the values assumed by $X_1,\ldots,X_n$. (For example $X_{1n}=\max(X_1,\ldots,X_n)$.)

Smirnov (1949) has shown that the class of all proper limit distributions for normalized r.v. X_{kn} consists of the following:

$$\Phi_\alpha(x;k)=\begin{cases}0 & \text{if } x<0\,,\\ \exp\left(-x^{-\alpha}\right)\displaystyle\sum_{s=0}^{k-1}x^{-s\alpha}/s! & \text{if } x\ge 0\,;\end{cases}$$

$$\Psi_\alpha(x;k)=\begin{cases}\exp\left(-|x|^{\alpha}\right)\displaystyle\sum_{s=0}^{k-1}|x|^{s\alpha}/s! & \text{if } x<0\,,\\ 1 & \text{if } x\ge 0\,;\end{cases}$$

where $\alpha>0$.

The limit distributions for the maximal term are obtained by putting $k=1$. Smirnov has shown that the domain of attraction of any cdf above does not depend on k, i.e. it coincides with the domain of attraction of the corresponding cdf which is obtained by putting $k=1$.

Mejzler and Weissman (1969) generalized Smirnov's result for the case where the initial r.v.'s are not necessarily identically distributed.

Galambos (1978) has shown that exchangeability plays an important role for extreme values. He has extended Smirnov's result to the case of exchangeable variables. A typical non-trivial example of exchangeable sequence are random spacings:

If $X_1,X_2,\ldots$, is a sequence of i.i.d. r.v. uniformly distributed on [0,1], if $0=X_0^{(n)}<X_1^{(n)}<\cdots<X_{n-1}^{(n)}<X_n^{(n)}=1$ are the order statistics corresponding to $0,1,X_1,\ldots,X_{n-1}$, then the random spacings of order n are defined by

$$S_i^{(n)} = X_i^{(n)} - X_i^{(n-1)}, \qquad i = 1, 2, \ldots, n,$$

and the maximal uniform spacing M'_n is defined by

$$M'_n = \max S_i^{(n)}.$$

The study of spacings has been the object of many publications (see, e.g., Pyke (1965), for general references). Levy (1939) has obtained its limit distribution:

$$\lim_{n\to\infty} P(nM'_n/\log n < x) = \exp(e^{-x}).$$

This result has been extended by a number of authors in the early '80s.

If $X_1, X_2, \ldots$ are i.i.d. random variables, $Y_n = \max(X_1, \ldots, X_n)$, and if $N(n)$ is a positive integer valued random variable, independent of $X_1, X_2, \ldots$, it is of interest to evaluate the distribution of $Y_{N(n)}$.

If it is assumed that $N(n)/n \to \tau$, $n \to \infty$, where τ is a (positive) random variable, and if there exist sequences $a_n > 0$ and b_n such that $(Y_n - b_n)/a_n$ converges weakly to a nondegenerate distribution function $H(\cdot)$ (belonging to one of the three basic types) then, as $n \to \infty$,

$$\lim_{n\to\infty} P((Y_{N(n)} - b_n)/a_n < x) = \int_0^{+\infty} H^y(x)dP(\tau < y).$$

Extensions can be made to the kth extremes in a similar manner. First results in this field have been obtained, among others, by Barndorff-Nielsen (1964).

Finally it should be noted that the maximum deviation between density estimates and the density has (when the density is smooth enough) a limiting type 1 (Gumbel) distribution.

In view of the importance of the limiting distributions discovered by Gnedenko, we conclude these comments by providing a summary and several examples of determining from the limiting type of an extremal distribution the given the initial distribution (the so-called "domain of attraction" introduced in Sec. 1.3).

Summary of Univariate Extreme Value Limiting Distributions

For Maxima (Standardized Form)

Gumbel (Type 1) $H_1(x) = \exp[-\exp(-x)^k], \qquad -\infty < x < \infty,$

Fréchet (Type 2) $H_2(x) = \begin{cases} \exp((-x)^{-k}) & x > 0, \\ 0 & x \le 0 \end{cases}$

Weibull (Type 3) $H_3(x) = \begin{cases} 1 & x \ge 0 \\ \exp[-(-x)^k] & x < 0 \end{cases}$

For Minima (Standardized Form)

Gumbel (Type 1) $L_1(x) = 1 - \exp[-\exp(-x)], \qquad -\infty < x < \infty,$

Fréchet (Type 2) $L_2(x) = \begin{cases} 1 - \exp(-(-x)^{-k}) & x > 0\,, \\ 1 & x \geq 0 \end{cases}$

Weibull (Type 3) $L_3(x) = \begin{cases} 1 - \exp(-x^k) & x > 0 \\ 0 & x \leq 0\,. \end{cases}$

(The Fréchet type is sometimes referred to as the *Cauchy–Fréchet* type.)

For the initial uniform distribution $F(x) = x$, $0 \leq x \leq 1$, it follows immediately by checking the validity of the condition presented in Sec. 1.3 that the limiting distribution for maxima in this case is the type 3 (Weibull) distribution.

For the initial Cauchy distribution, $F(x) = (1/2) + (\arctan x/\pi)$, $-\infty < x < \infty$, direct calculations applied to the condition presented in Sec. 1.3 show that the limiting distribution for the maxima in this case is the Fréchet distribution and by symmetry the same conclusion holds for the minima.

Similarly, recalling that for an exponential cdf $F(x) = 1 - \exp(-x/\lambda)$, $x > 0$, the percentiles $x_{1-1/n}$ are $-a\log(1/n)$ and $x_{1-(ne)^{-1}} = a\log(ne)$ respectively, after some simple calculations we arrive at

$$\lim_{n\to\infty} n\{\exp[-(\log n - x)]\} = e^{-x}\,.$$

This shows that the limiting distribution for maxima for the initial exponential distribution is the type 1 (Gumbel) distribution. The fact that for the initial exponential distribution, the limiting extreme value distribution is of type 1 is easily deduced from von Mises' (1936) sufficient conditions recalling that for this distribution the hazard rate $r(x) = f(x)/1 - F(x)$ is constant. In Table A.1 we summarize the forms of limiting distributions for maxima and minima for seven most widely used continuous distributions.

Table A.1:

Initial Distribution	Limiting Distribution for Extremes	
	Maxima	Minima
1. Exponential	Type 1 (Gumbel)	Type 3 (Weibull)
2. Gamma	Type 1 (Gumbel)	Type 3 (Weibull)
3. Normal	Type 1 (Gumbel)	Type 1 (Gumbel)
4. Log-normal	Type 1 (Gumbel)	Type 1 (Gumbel)
5. Uniform	Type 3 (Weibull)	Type 3 (Weibull)
6. Pareto	Type 2 (Fréchet)	Type 3 (Weibull)
7. Cauchy	Type 2 (Fréchet)	Type 2 (Fréchet)

B. Dependent Variables

Buishand (1985), among others, investigated the limiting distribution of maxima of sequences of dependent random variables. He points out that the classical results related to type 1 Gumbel distribution remain valid if the sequence $X_1, X_2, \ldots$ is a *mixing* sequence. This condition requires that: (1) the various terms in the sequence are "weakly dependent when their separation is large". For example in the mixing sequence

$$P(X_1 < x, X_2 < x, X_k < x) \to P(X_1 < x, X_2 < x) \cdot P(X_k < x)$$

as $k \to \infty$.

(2) $P(X_{i+k} \geq r | X_i \geq r) \to 0$ as $x \to \infty$ for every $k \neq 0$ (the right tail asymptotic independence). This implies no local clustering of exceedances of a high-level x. Details are given in Galambos (1987, Chap. 3) and Leadbetter *et al.* (1983, Chap. 3). For sequences of 1-*dependent random variables* namely for which the events $\{X_1 < x_1, \ldots, X_j < x_j\}$ and $\{X_{j+k} < x_{j+k}, \ldots, X_n < x_n\}$ are dependent for $k = 1$ and are independent for $k > 1$, the limiting distribution of $Y_n = \max(X_1, \ldots, X_n)$ may not coincide with the Gumbel distribution.

Greig (1967) provides an illuminating example involving the normal distribution.

For m-dependent sequences (where the events are independent if they are separated by more than m units), $m > 1$, it is possible that exceedances of high level x may occur in runs and also the runs may occur in "bunches". In that case the asymptotic Gumbel distribution is also not valid. Specifically for the distribution of the maximum, the number of clusters rather than the number of runs of individual exceedances has to be taken into account. (See, e.g., Rootzèn (1978).) This is especially relevant for the extreme value distribution of rainfall data (Buishand (1985); Marshall (1983)).

Chapter 2

Generalized Extreme Value Distributions

2.1 Basic Properties

The generalized extreme value (GEV) distribution was first introduced by Jenkinson (1955). The cumulative distribution function of the generalized extreme value distributions is given by

$$F_X(x) = \begin{cases} e^{-(1+\xi((x-\mu)/\sigma))^{-1/\xi}} & -\infty < x \leq \mu - \sigma/\xi \quad \text{for } \xi < 0; \\ & \mu - \sigma/\xi \leq x < \infty \quad \text{for } \xi > 0; \\ e^{-e^{-(x-\mu)/\sigma}} & -\infty < x < \infty \quad \text{for } \xi = 0\,. \end{cases} \tag{2.1}$$

The distribution above includes the type 2 distribution in Eq. (1.2) when $\xi > 0$, the type 3 distribution in Eq. (1.3) when $\xi < 0$, and the type 1 distribution in Eq. (1.1) when $\xi = 0$. The distribution is also referred to as the *von Mises type extreme value distribution* or the *von Mises–Jenkinson type distribution.*

The density function corresponding to (2.1) is

$$p_X(x) = \begin{cases} e^{-(1+\xi(x-\mu)/\sigma))^{-\frac{1}{\xi}}} \dfrac{1}{\sigma}\left\{1+\xi\left(\dfrac{x-\mu}{\sigma}\right)\right\}^{-\frac{1}{\xi}-1} & \\ & -\infty < x \leq \mu - \dfrac{\sigma}{\xi} \quad \text{for } \xi < 0\,; \\ & \mu - \dfrac{\sigma}{\xi} \leq x < \infty \quad \text{for } \xi > 0\,; \\ e^{-e^{-(x-\mu)/\sigma}} \dfrac{1}{\sigma} e^{-(x-\mu)/\sigma} & -\infty \leq x < \infty \quad \text{for } \xi = 0\,. \end{cases} \tag{2.2}$$

The standard form of the generalized extreme value distributions has the cdf

$$F_Y(y) = \begin{cases} e^{-(1+\xi y)^{-\frac{1}{\xi}}} & -\infty < y \le -1/\xi \quad \text{for } \xi < 0; \\ & -\dfrac{1}{\xi} \le y < \infty \quad \text{for } \xi > 0; \\ e^{-e^{-y}} & -\infty < y < \infty \quad \text{for } \xi = 0 \end{cases} \tag{2.3}$$

and the pdf

$$p_Y(y) = \begin{cases} e^{-(1+\xi y)^{-\frac{1}{\xi}}}(1+\xi y)^{-\frac{1}{\xi}-1} & -\infty < y \le -\dfrac{1}{\xi} \quad \text{for } \xi < 0; \\ & -\dfrac{1}{\xi} \le y < \infty \quad \text{for } \xi > 0; \\ e^{-e^{-y}}e^{-y} & -\infty < y < \infty \quad \text{for } \xi = 0. \end{cases} \tag{2.4}$$

We shall often use the abbreviation GEV (μ, σ, ξ).

The parameter ξ is called the shape parameter and may be used to model a wide range of tail behavior. The case $\xi > 0$ is that of a polynomially decreasing tail function and therefore corresponds to a long-tailed parent distribution. The case $\xi = 0$ is that of an exponentially decreasing tail, while $\xi < 0$ is the case of a finite upper endpoint and is therefore short-tailed.

Maritz and Munroe (1967) studied order statistics from this generalized extreme value distribution, and presented tables of means of order statistics from sample sizes 5 to 10 for the choices of the shape parameter $\xi = -0.4(0.05)0.10$. These authors have also discussed the estimation of all three parameters by the use of order statistics.

There are a number of non-regular situations associated with ξ: when $\xi < -1$ the maximum likelihood estimates do not exist, when $-1 < \xi < -1/2$ there may be problems, and when $\xi > 1/2$ the second and higher moments do not exist. A recent method proposed by Castillo and Hadi (1997) circumvents some of these problems: it provides well-defined estimates for all parameter values and performs well compared to any of the existing methods. (Fortunately, the experience with real-world data suggests that the condition $-1/2 < \xi < 1/2$ is almost always satisfied in practical applications – in particular in environmetrics.)

Suppose $X_1, X_2, \ldots$ are i.i.d. random variables with common cdf $F \in D(G)$ where G is GEV (μ, σ, ξ) for some μ, σ and ξ. Let $M_n^{(i)}$ denote the ith largest of the first n random variables, $i = 1, 2, \ldots, r$. The limiting joint distribution for the r largest order statistics for $x_1 \ge x_2 \ge \cdots \ge x_r$ is:

$$\Pr\left\{\frac{M_n^{(1)} - b_n}{a_n} < x_1, \frac{M_n^{(2)} - b_n}{a_n} < x_2, \ldots, \frac{M_n^{(r)} - b_n}{a_n} < x_r\right\}$$

$$\to \sum_{s_1=0}^{1}\sum_{s_2=0}^{2-s_1}\cdots\sum_{s_{r-1}=0}^{r-1-s_1-\cdots-s_{r-2}} \frac{(\gamma_2-\gamma_1)^{s_1}}{s_1!}\cdots\frac{(\gamma_r-\gamma_{r-1})^{s_r-1}}{s_{r-1}!}\exp(-\gamma_r). \tag{2.5}$$

Here $\gamma_i = -\log G(x_i; 0, 1, \xi)$ and $\{a_n\}$, $\{b_n\}$ are normalizing constants. (The related problem of the joint limiting distribution $(n-r)$ smallest order statistics was also considered.)

Practical applications of (2.5) proceed by assuming that n is sufficiently large for the limit law to hold exactly. We can also express the joint density of $(M_n^{(1)}, M_n^{(2)}, \ldots, M_n^{(r)})$ in the generalized form:

$$f(x_1, x_2, \ldots, x_r) = \sigma^{-r} \exp\left[-\left\{ 1 + \xi\left(\frac{x_r - \mu}{\sigma}\right)\right\}^{-1/\xi} \right.$$
$$\left. - \left(\frac{1}{\xi} + 1\right) \sum_{j=1}^{r} \log\left\{ 1 + \xi\left(\frac{x_j - \mu}{\sigma}\right)\right\}\right] \tag{2.6}$$

valid over the range $x_1 > x_2 \geq \cdots \geq x_r$ such that $1 + \xi(x_j - \mu)/\sigma > 0$ for $j = 1, 2, \ldots, r$. For the asymptotic approximation of (2.5) to be valid, r has to be small by comparison to n. As r increases, the rate of convergence of the limiting joint distribution decreases sharply. The choice of r is therefore crucial. Wang (1995) proposed a method for selecting r based on a suitable goodness-of-fit statistic.

From Eqs. (2.3) and (2.4), we deduce the characterizing differential equation

$$(1 + \xi y)p_Y(y) = -F_Y(y)/\log F_Y(y)\,. \tag{2.7}$$

Balakrishnan, Chan, and Ahsanullah (1993) have exploited the differential equation (2.7) to establish recurrence relations satisfied by the single and the product moments of lower record values. See also Ahsanullah (1994).

2.2 Statistical Inference (Classical Approach)

Ahsanullah and Holland (1994) have discussed the estimation of the location and scale parameters of the generalized extreme value distribution (when ξ is known) based on the record values.

The maximum likelihood estimation of the parameters μ, σ and ξ have been studied by a number of authors including Jenkinson (1969), Prescott and Walden (1980, 1983), Hosking (1985), and Macleod (1989). Based on a complete sample of size n from the generalized extreme value distribution (2.1), the Fisher expected information matrix is given by [Prescott and Walden (1980)]:

$$\begin{aligned}
E\left[-\frac{\partial^2 \log L}{\partial \mu^2}\right] &= \frac{n}{\sigma^2} p\,,\\
E\left[-\frac{\partial^2 \log L}{\partial \sigma^2}\right] &= \frac{n}{\sigma^2 \xi^2} - \{1 - 2\Gamma(2+\xi) + p\}\,,\\
E\left[-\frac{\partial^2 \log L}{\partial \xi^2}\right] &= \frac{n}{\xi^2}\left\{\frac{\pi^2}{6} + \left(1 - 0.5772157 + \frac{1}{\xi}\right)^2 - \frac{2q}{\xi} + \frac{p}{\xi^2}\right\},\\
E\left[-\frac{\partial^2 \log L}{\partial \mu \partial \sigma}\right] &= -\frac{n}{\sigma^2 \xi}\{p - \Gamma(2+\xi)\}\,,\\
E\left[\frac{\partial^2 \log L}{\partial \mu \partial \xi}\right] &= \frac{n}{\sigma \xi}\left(q - \frac{p}{\xi}\right),\\
E\left[\frac{\partial^2 \log L}{\partial \sigma \partial \xi}\right] &= \frac{n}{\sigma \xi^2}\left[1 - 0.5772157 + \frac{\{1 - \Gamma(2+\xi)\}}{\xi} - q + \frac{p}{\xi}\right],
\end{aligned} \tag{2.8}$$

where

$$p = (1+\xi)^2\Gamma(1+2\xi), \quad q = \Gamma(2+\xi)\left\{\psi(1+\xi) + \frac{1+\xi}{\xi}\right\},$$

and, as above, $\psi(\cdot)$ is the digamma function ($\psi(r) = d\log\Gamma(r)/dr$).

As mentioned above, the regularity conditions are satisfied when $\xi > -1/2$ and in this case the asymptotic variances and covariances of the maximum likelihood estimators are given by the elements of the inverse of the Fisher information matrix whose elements are presented above.

Hosking (1985) has provided a FORTRAN subroutine MLEGEV that facilitates the calculation of the maximum likelihood estimates of the parameters μ, σ and ξ (by the Newton–Raphson method) and the variance–covariance matrix of the estimated parameters. Macleod (1989) has suggested an adjustment that should be applied to Hosking's algorithm.

Otten and van Montfort (1980) consider the generalized extreme-value distribution of the form

$$F_X(x) = \Pr(X \le x) = \begin{cases} \exp\left[-(1-\theta z)^{1/\theta}\right] & 1/\theta < z < +\infty \quad \text{for } \theta < 0; \\ & -\infty < z < 1/\theta \quad \text{for } \theta > 0; \\ \exp\left[-\exp(-z)\right] & -\infty < z < +\infty \quad \text{for } \theta = 0. \end{cases}$$

(the parameter ξ is replaced here by $-\theta$).

Here $z = (x-\mu)/\sigma$; μ is a location parameter with $-\infty < \mu < +\infty$; σ is a scale parameter with $0 < \sigma < +\infty$ and θ is a shape parameter with $-\infty < \theta < +\infty$.

The inverse $x = F^{-1}(p)$ is given by:

$$x = \begin{cases} \mu + \sigma[1 - \exp(-\theta y)]/\theta, & \theta \ne 0 \\ \mu + \sigma y, & \theta = 0, \quad \text{where} \quad y = -\log\left[-\log(p)\right], \quad 0 < p < 1. \end{cases}$$

Note that the e^{-1}-point of this distribution is μ for any σ and θ.

Given a sample of n observations $X_1, X_2, \ldots, X_n$, denote $t = \theta z$ and

$$y = -\theta^{-1}\log(1-t) = z(1 + t/2 + t^2/3 + \cdots) \quad \text{(for small } t\text{)}.$$

Then $F(x) = \exp[-\exp(-y)]$. Observe that $t < 1$ unless $x = \mu + \sigma\theta$ which is the finite bound of the support of the distribution. The log (density) is given by

$$-\log(\sigma) - (1-\theta)y - \exp(-y),$$

and the log(likelihood) is:

$$L = \sum_{i=1}^{n}[-\log(\sigma) - (1-\theta)y_i - \exp(-y_i)].$$

To maximize L, Otten and van Montfort (1980) utilized the vector of first derivatives L' and the matrix of second derivatives L'' with respect to the parameters. For numerical calculations, for small t, series expansions of y and the derivatives y_θ and $y_{\theta\theta}$ are required. We shall briefly sketch the procedure.

Let $\mathbf{y}' = (y_\mu, y_\sigma, y_\theta)$. Here $y_\mu = \frac{-1}{\sigma(1-t)}$, $y_\sigma = \frac{-z}{\sigma(1-t)}$, $y_\theta = \frac{z}{\theta(1-t)} - \frac{y}{\theta}$. Then

$$(-1+\theta+e^{-y})\mathbf{y}' + \begin{pmatrix} 0 \\ -1/\sigma \\ y \end{pmatrix}$$

is the contribution of an observation to L'.

Let $\hat{\pi}_j$ be the jth estimate of the parameter $\pi = (\mu, \sigma, \theta)^{\mathrm{T}}$. Then, in the obvious notation,

$$\hat{\pi}_{j+1} = \hat{\pi}_j + (-L_j'')^{-1} L_j' \,, \tag{2.8a}$$

and, in the case of convergence, the maximum likelihood estimator $\hat{\pi} = \hat{\pi}_\infty$. The anticipated improvement when passing from $\hat{\pi}_j$ to $\hat{\pi}_{j+1}$ can break down if L'' has negative eigenvalues.

The starting value π_0 is the ML-estimator of μ and σ for $\theta = 0$ (the type 1 extreme value distribution).

A popular stopping rule is determined by:

$$\sum_j [(\hat{x}_p)_{j+1} - (\hat{x}_p)_j]^2 / [x_{(n)} - x_{(1)}]^2$$

with $p = 0.01$, 0.50, 0.99, $\hat{x}_p$ being the estimated p-point of the distribution, and stopping at, e.g., 10^{-8}.

Otten and van Montford (1988) recommended halving the first correction when L is not well approximated by a second-degree polynomial in the neighborhood of $\hat{\pi}_0$.

Table of approximate values of $(-L_j'')^{-1}$ was provided by Jenkinson (1969) and reproduced in *Flood Studies Report* (1975). However, Otten and van Montfort recommended working without approximations. As mentioned above, regularity conditions are satisfied for $\theta < 1/2$, and in this case asymptotic variances and covariances are given by the elements of L''^{-1} ($-E(L'')$ exists only for $\theta < 1/2$). Details of application of the iterative procedure (2.8a) are provided in a technical report by Prescott and Walden, University of Southampton, England (1985), and in Prescott and Walden (1980) discussed above.

Hosking, Wallis, and Wood (1985) have also discussed the method of probability-weighted moments (PWM) for the estimation of the parameters μ, σ and ξ. What the PWM estimators would seem to have in their favor is that their evaluation is simple and guaranteed for ξ in the range $[-1/2, 1/2]$ whereas convergence of the maximum likelihood estimates is erratic for ξ close to $-1/2$. As described in Chap. 1 in this approach, one considers the moments

$$\beta_r = E[X\{F(X)\}^r], \qquad r = 0, 1, 2, \ldots\,, \tag{2.9}$$

and sets up the necessary number of moment equations by using the sample statistics

$$b_r = \frac{1}{n} \sum_{i-1}^{n} \frac{(i-1)^{(r)}}{(n-1)^{(r)}} X_i' \,, \qquad r = 0, 1, 2, \ldots\,, \tag{2.10}$$

which are unbiased estimators of the moments β_r. One may instead use the simplified estimates

$$\hat{\beta}_r[p_{i,n}] = \frac{1}{n} \sum_{i=1}^{n} p_{i,n}^r X_i' \,, \tag{2.11}$$

where $p_{i,n}$ is a plotting position [a distribution-free estimate of $F(X_i')$] that may be taken as

$$p_{i,n} = \frac{i-a}{n} \,, \qquad 0 < a < 1 \,,$$

or

$$p_{i,n} = \frac{i-a}{n+1-2a}, \qquad -\frac{1}{2} < a < \frac{1}{2}.$$

For the generalized extreme value distribution. Hosking *et al.* (1985) derived

$$\beta_r = (r+1)^{-1}\left[\mu - \frac{\sigma}{\xi}\left\{1 - \frac{\Gamma(1-\xi)}{(1+r)^{-\xi}}\right\}\right], \qquad \xi < 1, \quad \xi \neq 0. \tag{2.12}$$

They used (2.12) to show the following relations between the simplified estimators needed to determine an estimator of ξ.

$$\hat{\beta}_0 = \beta_0 = \mu - \frac{\sigma}{\xi}\{1 - \Gamma(1-\xi)\}, \tag{2.13}$$

$$2\hat{\beta}_1 - \hat{\beta}_0 = 2\beta_1 - \beta_0 = -\frac{\sigma}{\xi}\Gamma(1-\xi)(1-2^{\xi}), \tag{2.14}$$

and

$$\frac{3\hat{\beta}_2 - \hat{\beta}_0}{2\hat{\beta}_1 - \hat{\beta}_0} = \frac{3\beta_2 - \beta_0}{2\beta_1 - \beta_0} = \frac{1-3^{\xi}}{1-2^{\xi}}. \tag{2.15}$$

The exact solution for ξ from Eq. (2.15) requires iterative methods. However, since the function $(1-3^{\xi})/(1-2^{\xi})$ is almost linear over the range of interest $(-1/2 < \xi < 1/2)$, the following approximate low-order polynomial estimator is used:

$$-\hat{\xi} = 78590c + 2.9554c^2, \tag{2.16}$$

where

$$c = \frac{2\hat{\beta}_1 - \hat{\beta}_0}{3\hat{\beta}_2 - \hat{\beta}_0} - \frac{\ln 2}{\ln 3}.$$

The error due to using (2.16) is less than 0.0009 throughout the range $-1/2 < \xi < 1/2$ (Hosking *et al.*, 1985). Given $\hat{\xi}$, the scale and location parameters can be estimated successfully as:

$$\hat{\sigma} = -\frac{(2\hat{\beta}_1 - \hat{\beta}_0)\hat{\xi}}{\Gamma(1-\hat{\xi})(1-2^{\hat{\xi}})}, \qquad \hat{\mu} = \hat{\beta}_0 - \frac{\hat{\sigma}}{\hat{\xi}}\{\Gamma(1-\hat{\xi}) - 1\}. \tag{2.17}$$

The PWM estimate of the bound is thus $\hat{\mu} - \hat{\sigma}/\hat{\xi}$, where $\hat{\mu}$, $\hat{\sigma}$ and $\hat{\xi}$ are as in (2.16) and (2.17). Note, however, that the use of PWM estimators does not guarantee that

$$x_i \leq \hat{\mu} - \frac{\hat{\sigma}}{\hat{\xi}}, \quad \forall x_i, \quad i = 1, \ldots, n, \quad \text{for} \quad \hat{\xi} < 0$$

or

$$x_i \geq \hat{\mu} - \frac{\hat{\sigma}}{\hat{\xi}}, \quad \forall x_i, \quad i = 1, \ldots, n, \quad \text{for} \quad \hat{\xi} > 0.$$

Consequently, PWM (like the Method of Moments) have no built-in feature to ensure feasibility and can yield non-feasible parameters estimates. For given μ, σ, ξ and n, we are interested in the probability that PWM estimators are not feasible and wish to compute

$$\begin{aligned} P\left(\max x_i > \hat{\mu} - \frac{\hat{\sigma}}{\hat{\xi}}\right) &= 1 - P\left(\text{all } x_i \text{ from a sample of size } n < \hat{\xi} - \frac{\hat{\sigma}}{\hat{\xi}}\right) \\ &= 1 - \left\{P\left(X < \hat{\mu} - \frac{\hat{\sigma}}{\hat{\xi}}\right)\right\}^n, \end{aligned} \tag{2.18}$$

where $X \sim$ GEV (μ, σ, ξ). (We are assuming that ξ is negative.) Simulations showed that the probability (2.18) has the following properties: (1) for a fixed n, it decreases with ξ and (2) for a fixed ξ it increases and then decreases as $n \to \infty$.

Consider the random vector $(B_0, B_1, B_2)^{\mathrm{T}}$ where realizations β_0, β_1, β_2 are

$$\beta_0 = \frac{1}{n}\sum_{i=1}^{n} x_i\,, \qquad \beta_1 = \frac{1}{n}\sum_{i=1}^{n} \frac{i-1}{n-1} x_{i:n}\,, \qquad \beta_2 = \frac{1}{n}\sum_{i=1}^{n} \frac{(i-1)(i-2)}{(n-1)(n-2)} x_{i:n} \tag{2.19}$$

and $x_{1:n} \le \cdots \le x_{n:n}$ is the ordered sample. Then the upper bound $\hat{\mu} - \hat{\sigma}/\hat{\xi}$ is a realization of the random variable

$$g(B_0, B_1, B_2) = B_0 + (2B_1 - B_0)/(1 - 2^{\hat{\xi}})\,,$$

where $\hat{\xi}$ is given in (2.16).

We seek an estimate of the mean and the variance of $g(B_0, B_1, B_2)$. The distribution of $g(B_0, B_1, B_2)$ is quite involved and Dupuis (1996) obtained an approximation by means of simulation in terms of a lognormal random variable.

While there is no analytical solution to Dupuis' (1996) approximation (expressed in a form of an integral), the integrand is well-behaved and numerical integration provides very accurate results. Numerical results of Dupuis (1996) investigations show that nonfeasibility is less than 1% for all n when $\xi > 0$. Nonfeasibility is also low for large negative ξ and it is only for $\xi \le -0.2$ that we observe an appreciable nonfeasibility as high as 20%. Also, for a fixed ξ, the probability of obtaining PWM estimates which are not feasible increases with n, before eventually decreases to 0. The author strongly advises practitioners to use the more numerically intensive and difficult maximum likelihood procedure to assure feasibility.

Chen and Balakrishnan (1995) also observed that the PWM method when estimating parameters of the generalized extreme value distribution can lead to infeasible estimates (in the sense that the estimated distribution has an upper bound and one or more of the data values lie outside this bound). The authors propose a modification which alleviates this problem except for small sample sizes ($n = 15$ or 25) when — based on their calculations — probabilities of obtaining infeasible parameter estimates are almost always greater than 5%.

Hosking (1986) also noticed this problem and suggested the following *ad hoc* method to overcome the difficulty.

Let x denote $x_{1:n}$ or $x_{n:n}$; if the boundary condition is found to be violated by the PWM estimators of the parameters, he recommends equating x to $\mu - \frac{\sigma}{\xi}$ and solving for ξ. This leads to $\hat{\xi} = \ln\{(2b_1 - x)/(b_0 - x)\}/\ln 2$ for the generalized extreme value distribution and the other parameters are estimated as before. Recall that b_r are defined in (2.10).

Using standard arguments, Hosking *et al.* (1985) have shown that the asymptotic variance–covariance matrix of PWM estimates $(\hat{\mu}, \hat{\sigma}, \hat{\xi})^{\mathrm{T}}$ is given by

$$\frac{1}{n}\begin{bmatrix} \sigma^2 w_{11} & \sigma^2 w_{12} & \sigma w_{13} \\ & \sigma^2 w_{22} & \sigma w_{23} \\ & & w_{33} \end{bmatrix}, \tag{2.20}$$

where the w's depend only on ξ. The asymptotic efficiency of the individual PWM estimators and the overall efficiency are presented in Fig. 2.1 [from Hosking *et al.* (1985)].

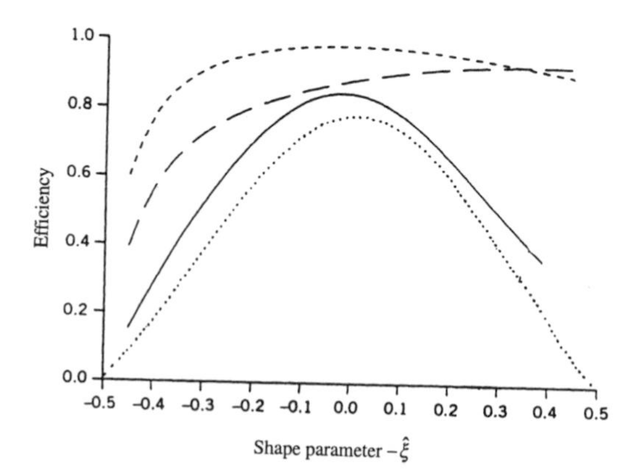

Figure 2.1: Asymptotic efficiency of PWM estimators of parameters of the GEV distribution: —— $\hat{\xi}$; - - - - $\hat{\sigma}$; - - - - - - $\hat{\mu}$; $\cdots$ overall efficiency (ratio of the determinants of asymptotic covariance matrices of ML and PWM estimators).

In defining *partial* probability-weighted moments, Wang (1990) discussed the estimation of the parameters of the generalized extreme value distribution based on censored samples. Prescott and Walden (1983) have discussed the maximum likelihood estimation of the parameters based on a doubly Type II censored sample $X'_{r+1}, \ldots, X'_{n-s}$ (where the smallest r and the largest s observations are censored in a sample of size n) from the generalized extreme value distribution (2.1). They have also presented expressions for the asymptotic variance–covariance matrix of these MLEs.

T. E. Smith (1984) has discussed a choice probability characterization of generalized extreme value models. Testing whether the shape parameter ξ is zero in the generalized extreme value distributions for the data at hand was addressed by Hosking (1984). Some goodness-of-fit tests for the generalized extreme value distributions have been examined by Chowdhury *et al.* (1991). They have calculated critical points for the Kolmogorov–Smirnov test for the values of the shape parameter $\xi = -0.20(0.50)0.25$ and the scale parameter $\sigma = 0.01, 0.05$ and 0.10. An excellent discussion on the models for exceedances over high thresholds by Davison and Smith (1990) provides further insight into issues related to these distributions. By using a predictive likelihood that approximates both Bayesian and maximum likelihood predictive inference, Davison (1986) has applied it to the prediction of extremes by means of the generalized extreme value distribution.

2.3 Bayesian Inference

The distribution function for the generalized extreme value family can be written as

$$G(z) = \exp\left\{ -\left[1 + \xi\left(\frac{z-\mu}{\sigma}\right)\right]_{+}^{-1/\xi} \right\}. \tag{2.21}$$

The parameters μ and σ are location and scale parameters and ξ is a shape parameter determining the weight of the tail of G and thus of the initial distribution function F of the series $X_1, X_2, \ldots, X_n$. Recall that as $\xi \to 0$, we have the Gumbel (type 1) model with

$$G(z) = \exp\left[-\exp\left\{-\frac{z-\mu}{\sigma}\right\}\right].$$

Much of Bayesian modeling of extremes has focused on the latter family.

The series $M_n = \max(X_1, \ldots, X_n)$ is often restricted in application to annual maximum data while in fact the series $X_1, X_2, \ldots, X_n$ may contain other data informative about the tail of F. This leads to the distribution of "threshold exceedances"

$$Y_i = X_i - u | X_i > u$$

which is taken as

$$H(y) = 1 - (1 + \xi y/\tilde{\sigma})_+^{-1/\xi},$$

where $\tilde{\sigma} = \sigma + \xi(u - \mu)$ and μ, σ and ξ are the GEV parameters. If the interest is in the lower tail, we apply similar argument to the series

$$m_n = \min\{X_1, X_2, \ldots, X_n\}$$

leading to the asymptotic distribution of minima of the form

$$G(z) = 1 - \exp\left\{-\left[1 + \xi\left(\frac{z-\mu}{\sigma}\right)\right]_+^{-1/\xi}\right\}.$$

This model (which includes Weibull distribution in the case $\xi < 0$) is often used for modeling failure time data.

There are relatively few papers linking directly the themes of extreme value modeling and Bayesian inference. In the reliability literature, there are substantially more references, presumably because there are fewer conceptual problems in formulating assessment in this context. (See Coles and Powell (1996) for a review.)

The difficulty in utilizing the Bayesian approach in extreme value problems is that the value of additional prior information is likely to be substantial, while the plausibility of formulating this kind of prior knowledge may be questionable for extremal behavior.

When applying Bayesian methodology we ought to strive to employ Bayesian procedures as a means to incorporate genuine scientific belief in data analysis rather than use Bayesian approach as simply as a formal technical inferential device. Nevertheless, it is the second approach that is often taken in applications. Unfortunately, the model

$$G(z) = \exp\left\{-\left[1 + \xi\left(\frac{z-\mu}{\sigma}\right)\right]_+^{-1/\xi}\right\}$$

admits in general no conjugate priors. In the restricted situation of a *single* parameter case: location parameter (when $\xi = 0$) or scale parameter, Engelund and Rackwitz (1992) obtained conjugate priors. The drawback here is that ξ is the most restrictive parameter. Ashour and El-Adl (1980) considered the distribution of the minima

$$G(z) = 1 - \exp\left\{-\left[1 + \xi\left(\frac{z-\mu}{\sigma}\right)\right]_+^{-1/\xi}\right\}$$

in the case when $\xi \to 0$, i.e. when the data is left-censored. In this case the joint conjugate prior is

$$f(\mu,\sigma) = \sigma^{-H} \exp\left[-\frac{G}{\sigma} + \frac{H\mu}{\sigma} - \frac{D\exp(\mu-\delta)}{\sigma}\right]$$

with parameters D, G, H and δ. Based on a simulated data example comparison between Bayesian and maximum likelihood approaches in this case, one may conclude that Bayesian estimators are more efficient but possess substantial bias.

Pickands (1994) suggested non-conjugate prior for the model (already mentioned above)

$$H(y) = 1 - (1+\xi y/\tilde{\sigma})_+^{-1/\xi} \quad \text{(which is the distribution of } Y_i = X_i - u|X_1 > u)$$

given by $f(\tilde{\sigma},\xi) \propto \frac{1}{\tilde{\sigma}}\mathbf{1}_{\{\tilde{\sigma}>0\}}$ (this is equivalent to specifying priors for $\log(\tilde{\sigma})$ and ξ that are independent and uniform on $(-\infty,\infty)$). Note, however, that $\tilde{\sigma}$ depends on the threshold u chosen in $Y_i = X_i - u|X_i > u$. Beirlant *et al.* (1999) recommend the so-called maximal data information prior: $\pi(\xi,\tilde{\sigma}) = \frac{1}{\tilde{\sigma}}e^{-(1+\xi)}$.

Achcar *et al.* (1987) modeled a sequence of survival times $T_1, T_2, \ldots, T_n$ where Y has the distribution of the minima:

$$G(z) = 1 - \exp\left\{-\left[1+\xi\left(\frac{z-\mu}{\sigma}\right)\right]_+^{-1/\xi}\right\}$$

and

$$Y = \begin{cases} \dfrac{T^\lambda - 1}{\lambda} & \text{for } \lambda \neq 0 \\ \log T & \text{for } \lambda = 0 \end{cases}$$

(the well-known Box–Cox transformation). Using the Jeffrey priors for μ and σ and an improper uniform prior for the Box–Cox parameter λ, they obtain the posterior marginal distribution for λ and assess the quality of the transformation by a simple probability plot.

Coles and Powell (1995) modified the Archcar *et al.* procedure (using their data).

In their study the maximum likelihood approach has failed due to clustering of data close to the start of the experiment causing unboundedness of the likelihood function. In principle, however Bayesian analysis is not affected by unboundedness – only complexity of computations increases. Coles and Powell (1995) chose almost flat prior distribution; marginal posterior distributions for μ, σ and ξ are presented in Fig. 2.2.

Figure 2.2(c) shows that $\xi < -1$ almost certainly. Skewness in the posterior densities may indicate that the posterior mean is outside the parameter space, namely there is a value of observed failure time $t < \bar{\mu} + \bar{\sigma}/\bar{\xi}$ for at least one t. Coles and Powell (1995) chose an estimator corresponding to the modes of the marginal distributions:

$$(\tilde{\mu},\tilde{\sigma},\tilde{\xi}) = (84.3, 151.0, -1.77).$$

Based on this estimator, the transformation

$$Y = 1/\xi \log\left(1 - \xi\left(\frac{T-\mu}{\sigma}\right)\right)$$

ought to produce variables with standard Gumbel distribution for minima. Standardized order statistics $y_{(i)}$ calculated from the original order statistics $t_{(i)}$ show, by means of a probability plot, that the GEV model is indeed adequate in this case.

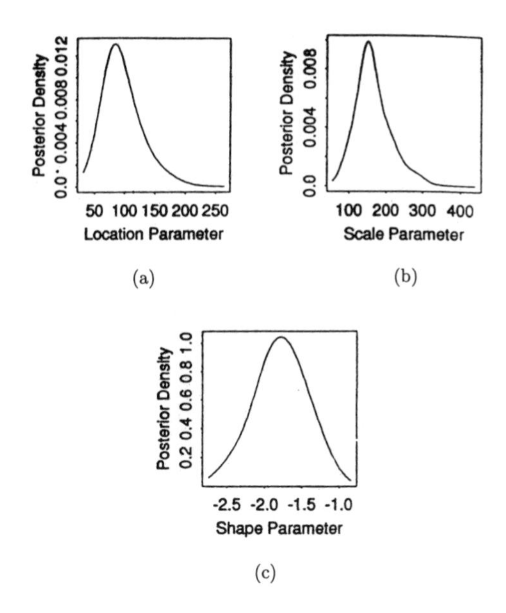

Figure 2.2: Posterior distributions of GEV parameters. (Coles and Powell (1995))

Predictive density function is quite appropriate for Bayesian analysis of extreme value distributions since in many applications the role of an extreme value analysis is to characterize the extremal behavior of the past history in order to be able to design against extreme excursions of future values.

For the GEV, Davison (1986) provided approximation to predictive density function

$$f(y|x) = \int f(y|\theta)f(\theta|x)d\theta$$

(here x is historical data, y is a future observation and $f(y|\theta)$ and $f(\theta|x)$ are respectively, the likelihood and posterior distribution of θ given x).

Engelund and Rackwitz (1992) — already mentioned above — calculated the exact form of predictive distributions for *one-parametric* cases of the general GEV model with the distribution function

$$G(z) = \exp\left\{-\left[1+\xi\left(\frac{z-\mu}{\sigma}\right)\right]_+^{-1/\xi}\right\}.$$

Specifically, let u be the location parameter in the extreme value Gumbel distribution

$$F_X(x|u) = \exp\left[-\exp\left[-\alpha(x-u)\right]\right].$$

Motivated by the fact that

$$r = \sum_{i=1}^{n} \exp\left(-\alpha x_i\right)$$

is a sufficient statistics for this parameter (see e.g. Schrupp and Rackwitz (1984)), Engelund and Rackwitz (1992) proposed the following family of priors:

$$f'(u) \propto \exp(p\alpha u)$$

where p is a constant. These priors are improper in the sense that they integrate to infinity. They correspond to noninformative priors (see, e.g., Box and Tiao (1962)) when $p = 0$. Note that even the first moment of these priors fails to exist. For type 1 (Gumbel) extreme value distribution, the predictive density function with this prior is

$$f_Y(x) = (n+p)\frac{\alpha}{r}\exp[-\alpha x]\left(1+\frac{\exp[-\alpha x]}{r}\right)^{-n-1-p}$$

and the cdf

$$F_Y(x) = \left(1+\frac{\exp[-\alpha x]}{r}\right)^{-n-p}.$$

See the figures below.

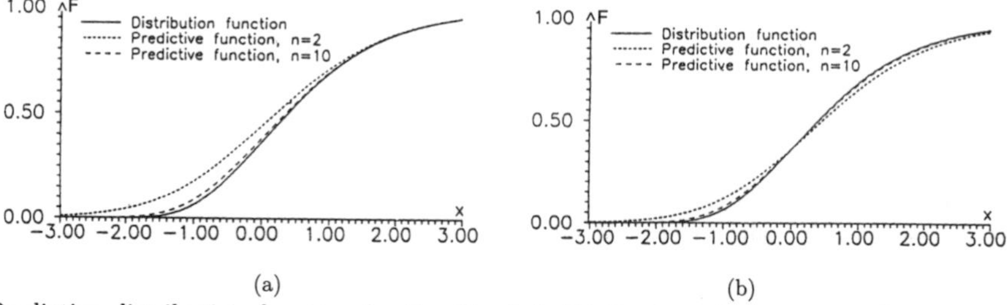

Predictive distribution function for the Gumbel distributions for two sets of parameters (α, u, p) and $n = 2, 10$. (a): (1.0, 0.0, 0.0); (b): (1.0, 0.0, 0.46).

It is seen that the predictive distribution function for the Gumbel distribution decreases as the number of experiments increases even for large values of $F_Y(\cdot)$. A similar behavior can be observed for the Fréchet distribution. On the contrary, for the Weibull distribution of minima, the predictive distribution function increases as the number of experiments increases.

Engelund and Rackwitz (1992) pointed out that this makes no sense if Gumbel (or Fréchet) distributions are used to model loads, while a Weibull model is adopted for resistance in structural reliability. They quote the well-known statement by Berger (1980):

> "... even unanimously acclaimed noninformative priors (such as those for location parameters or scale parameters) can lead to inferior decision rules."

Using the least informative prior, these authors succeeded in showing that such a choice corresponds to the value $p = 0.46$ and, in this case, the conjugate prior leads to a reasonable predictive decision as far as a function of the number of observations is concerned.

Lingappaiah (1984) investigated predictive probabilities of extreme order statistics under a sequential sampling scheme.

Smith and Naylor (1987) chose priors for the family

$$G(z) = 1 - \exp\left\{-\left[1+\xi\left(\frac{z-\mu}{\sigma}\right)\right]_+^{-1/\xi}\right\}$$

(the three-parameter Weibull distribution mentioned above) with $\xi < 0$ in an arbitrary manner so as to reflect a range of potential scientific hypotheses. The main purpose is to demonstrate the *computational* feasibility — within various extreme value models — of dealing with priors that may be analytically intractable but scientifically motivated.

Coles and Tawn (1996) express the opinion that it is unlikely that prior beliefs on extremal behavior could adequately be elicited directly in terms of the GEV parameters. Even having marginal priors for each parameter it is still unclear how to use them for construction of joint priors. One ought always remember that long range extrapolation is sensitive to the weight of the tail. These authors recommended eliciting prior information within a parametrization which corresponds to a scale on which the expert has familiarity and within which a natural dependence between the prior specifications is constructed. Starting from family (2.21) they invert the equation, to obtain the 1–p quantile of the annual maximum distribution:

$$q_p = \mu + \sigma[-\log(1-p)^{-\xi} - 1]\xi \tag{2.22}$$

and elicit prior information in terms of $(q_{p_1}, q_{p_2}, q_{p_3})$ for specified values of $p_1 > p_2 > p_3$. They choose the joint prior for q_{p_i} of the form

$$f(q_{p_1}, q_{p_2}, q_{p_3}) \propto q_{p_1}^{\alpha_1 - 1} \exp(-\beta_1 q_{p_1}) \prod_{i=2}^{3} (q_{p_i} - q_{p_{i-1}})^{\alpha_i - 1} \exp\{-\beta_i (q_{p_i} - q_{p_{i-1}})\}. \tag{2.23}$$

Substituting the quantile expression (2.22) into (2.23) and multiplying by the appropriate Jacobian of the transformation

$$(q_{p_1}, q_{p_2}, q_{p_3}) \to (\mu, \sigma, \xi)$$

leads to expression for the priors in terms of the GEV parameters. Multiplication by the appropriate likelihood gives the posterior distribution of (μ, σ, ξ). It turns out that in an example of extreme rainfall data (based on 54-year series of daily rainfall aggregates measured at a location in the Southwest England), analytical calculations of the marginal distributions are intractable. Nevertheless, the recently attained power and simplicity of Markov chain Monte–Carlo (MCMC) techniques suggest that direct simulation from a Markov chain whose equilibrium distribution is the prior π is straightforward. The authors' analysis is based on a Gibbs sampler, successively updating the individual parameters μ, σ and ξ from the prior π, conditionally on the current values of the other parameters. See Coles and Tawn (1996) for details. The results are summarized in Fig. 2.3. The authors observed that marginally, the location and scale parameter priors are almost non-informative: the prior for μ is very flat, whereas that for σ resembles $1/\sigma$. The marginal prior for ξ carries most information since we are dealing here with a distribution possessing an infinite upper end point. There is physical reasoning why the prior information for the location parameter should be highly diffused relative to that of the other parameters: the location parameter tends to be strongly dependent on localized site-specific characteristics which are difficult to calibrate without reference to data. In contrast, the scale (and, especially, the shape) parameters are governed

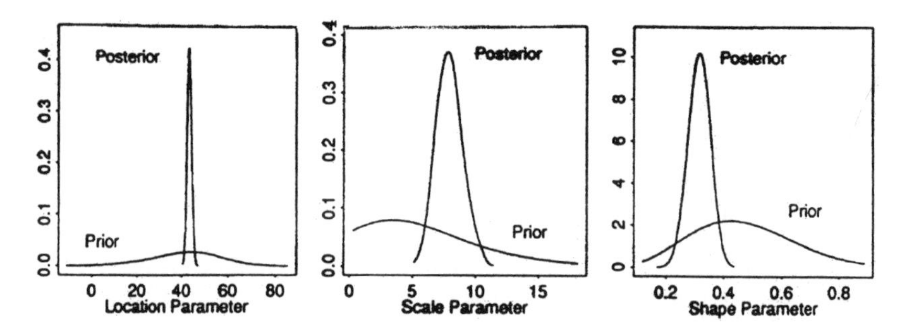

Figure 2.3: Univariate marginals of prior and posterior distributions of each of the GEV parameters. (Coles and Tawn (1996))

mainly by regional characteristics of a rainfall process, about which prior information is more easily assessed. (This reverses the situation from a likelihood-based analysis in which the relative precision of estimation of the location parameter is much greater than that of the scale or shape parameters.)

As to utility of Bayesian approach to the spatial modeling of extreme wind speeds, the observations and investigations due to Coles and Powell (1996) revealed that Bayesian analysis is quite suitable for spatial data exploration since it is quite reasonable to adopt the information acquired spatially as a prior belief, and, as data becomes available, we can either reinforce our faith in the prior or to modify our beliefs, accordingly which is in a harmony with the Bayesian paradigm. Their investigations of annual maximum wind speeds for 106 sites throughout the U.S.A. have shown that using the three-parameter GEV model relatively few data were required to calculated the posterior distribution of the location parameter μ, while the scale and shape parameters σ and ξ, respectively, necessitate much more data. As already mentioned above for short data records (5–10 years) maximum likelihood estimators are unreliable and possess low precision — having a tendency to oscillate as additional data become available — while the corresponding Bayesian estimates seem to be quite consistent with the estimates based on long-term data, having acceptable levels of precision and are by far more stable, especially when dealing with the shape parameter ξ. As it has already been mentioned this parameter affects significantly the long-term extrapolation. It should be observed that for "complete" data records Bayesian estimators are not inconsistent with maximum likelihood ones.

A special feature of Bayesian analysis of extreme-value data is related to the primary concern about the behavior *beyond* the range of the observed data, namely here the prior may fail to dominate data. Moreover, since the likelihood here is itself an asymptotic approximation, an optimal procedure may not be necessarily to include more data in the likelihood but perhaps to raise the threshold u in the distribution of "threshold exceedances".

2.4 Robust Estimation

Again consider the GEV distribution function H_θ involving a three-dimensional parameter

$$\boldsymbol{\theta} = [\mu, \sigma, \xi],$$

$$H_{\boldsymbol{\theta}} = H_{\mu,\sigma,\xi}(x) = \exp\left\{-\left(1 + \frac{\xi(x-\mu)}{\sigma}\right)^{-1/\xi}\right\};$$

here $1 + \frac{\xi(x-\mu)^{-1/\xi}}{\sigma} > 0$, $\sigma > 0$, but ξ and μ are arbitrary. (ξ is the shape parameter which as already mentioned above, in practice, usually lies in the range $-1/2 < \xi < 1/2$.) Both MLE and PWME of this parameter $\boldsymbol{\theta}$ have unbounded influence function and hence provide poor robustness behavior. For $\xi \neq 0$, the score function

$$s(x; \boldsymbol{\theta}) = (\partial/\partial\boldsymbol{\theta}) \log h_{\boldsymbol{\theta}}(x)$$

can easily be calculated explicitly. Here $h_{\boldsymbol{\theta}}(x)$ is the density corresponding to H_θ.

Given an estimator of $\boldsymbol{\theta}$, $T_n = T_n(x_1, x_2, \ldots, x_n)$, it can be viewed as a functional of the empirical distribution function $H^{(n)}$ (which puts mass $1/n$ on each observation); $T_n(x_1, \ldots, x_n) = T(H^{(n)})$. The basic tool to assess robustness is the influence function defined as:

$$\mathrm{IF}(x; T, H_{\boldsymbol{\theta}}) = \lim_{\xi \to 0} \frac{T((1-\varepsilon)H_{\boldsymbol{\theta}} + \varepsilon\Delta_x) - T(H_{\boldsymbol{\theta}})}{\varepsilon},$$

where Δ_x is a point mass at x. The IF describes the effect of a small contamination $\varepsilon\Delta_x$ at the point x on the estimate (standardized by the mass of the contamination). In fact, the linear approximation $\varepsilon\mathrm{IF}(x; T, H_\theta)$ measures the asymptotic bias of the estimator caused by the contamination. A desirable robustness property for an estimator is that it has a bounded IF. Such an estimator is called *B-robust* (*bias-robust*).

Dupuis and Field (1998) constructed Optimal Biased – Robust estimators (OBRE) and compared them with Probability Weighted Moment Estimators (PWME) originated by Hosking (1986).

Definition of OBRE is rather involved and is related to M-estimators, which are in turn a generalization of the maximum likelihood estimators.

An M-estimator is the solution T_n of the equation

$$\sum_{i=1}^{n} \boldsymbol{\psi}(x_i; T_n) = \mathbf{0}$$

for some function $\boldsymbol{\psi} : X \times R^p \to R^p$ ($p = 3$ in the case of GEV distributions), X is the space of observations, R^p is the space of parameters. There are several versions of the OBRE differing in the way they bound the influence function (IF). Dupuis and Field used the standardized OBRE for a given bound c (the robustness constant) on IF and the OBRE is *implicitly* defined by

$$\sum_{i=1}^{n} \boldsymbol{\psi}(x_i; \boldsymbol{\theta}) = \sum_{i=1}^{n} \{s(x_i; \boldsymbol{\theta}) - \mathbf{a}(\boldsymbol{\theta})\} W_c(x_i; \boldsymbol{\theta}) = 0, \tag{2.24}$$

where $s(x; \boldsymbol{\theta})$ is the score function and $W_c(x; \boldsymbol{\theta})$ is the weight function

$$W_c(x; \boldsymbol{\theta}) = \min\left\{1, \frac{c}{||A(\boldsymbol{\theta})\{s(x, \boldsymbol{\theta}) - \mathbf{a}(\boldsymbol{\theta})\}||}\right\},$$

$||\cdot||$ denotes the Euclidean norm. In turn, the $p \times p$ matrix $A(\boldsymbol{\theta})$ and the $p \times 1$ vector $\mathbf{a}(\boldsymbol{\theta})$ are defined implicitly by

$$E\{\psi(x; \boldsymbol{\theta})\psi(x; \boldsymbol{\theta})^{\mathrm{T}}\} = \{A(\boldsymbol{\theta})^{\mathrm{T}} A(\boldsymbol{\theta})\}^{-1},$$

$$E\psi(x; \boldsymbol{\theta}) = \mathbf{0}. \tag{2.25}$$

Dupuis and Field (1998) provided a detailed algorithm for OBRE estimation of parameters of GEV distribution. The basic idea is to start with the score function and modify it in such a manner that the influence function will be bounded and satisfy Eq. (2.25). Note that

$$\mathbf{a}(\boldsymbol{\theta}) = \frac{\int s(x,\boldsymbol{\theta})W_c(x,\boldsymbol{\theta})dH_{\boldsymbol{\theta}}(x)}{\int W_c(x,\boldsymbol{\theta})dH_{\boldsymbol{\theta}}(x)}$$

and

$$A^{\mathrm{T}}A = M_2^{-1},$$

where

$$M_k = \int \{s(x,\boldsymbol{\theta}) - \mathbf{a}(\boldsymbol{\theta})\}\{s(x,\boldsymbol{\theta}) - \mathbf{a}(\boldsymbol{\theta})\}^{\mathrm{T}} W_c(x,\boldsymbol{\theta})^k \, dH_{\boldsymbol{\theta}}(x), \qquad k = 1, 2\,.$$

The weight function W_c multiplies the score function $\mathbf{s}(x_i, \boldsymbol{\theta})$ and after subtracting $\mathbf{a}(\boldsymbol{\theta})$ it has been made to satisfy (2.24) (the so-called Fisher consistency). The weights will always be less than or equal to 1 and will be less than 1 when the norm of the appropriately standardized score function exceeds the cutoff c. This downweights observations which are not close to the working values of the parameters in the given score-function metric. For observations which are consistent with the current values of $\boldsymbol{\theta}$, the weights are one and the score function is that of the maximum likelihood estimate. This ensures that the OBRE is efficient, since it is as similar as possible to the MLE for the bulk of the data.

In a simulation experiment carried out by Dupuis and Field (1998), the most robust estimator was at $c = \sqrt{p}$ and the most efficient (the maximum likelihood one) was obtained at $c = \infty$. The constant c acts as regulator between robustness and efficiency. The authors also provide an illuminating example to illustrate the behavior of OBRE and contrast it with the PWM estimator. The data consist of 40 annual maximum one-day rainfalls at Fredericton, New Brunswick, Canada and correspond to the years 1951–1990. The estimates with their asymptotic standard deviations are as follows:

Table 2.1:

Method	μ	σ	ξ
OBRE ($c = 4$)	44.07(3.3)	15.29(2.9)	−0.146(0.13)
PWME	45.3(2.4)	13.1(1.9)	0.10(0.13)

As can be seen, the estimates of μ and σ are quite similar for the two methods and lie within 95% confidence intervals of the other estimate. On the other hand, the estimates of ξ differ substantially. It is also noteworthy that the sign of ξ has changed, indicating a lower bound of −146.8 for the OBRE and an upper bound of 85.7 for the PWME. If we look at the weights assigned by OBRE, we see that the largest observation, 146.8, has a weight of 0.00014, indicating that this observation is not well fitted by the GEV distribution. This point is reinforced by the quantile–quantile plot using the OBR estimates, where 146.8 clearly lies some distance from the least-median-of-squares line drawn through the data. The OBRE gives estimates very similar to the PWME when the data are well fitted by a GEV distribution, but OBRE also provides a good indication of the lack of fit through the weights. Also the OBRE tends to be more efficient than PWME for the model when $\xi > 0$ and less efficient when $\xi < 0$ which corresponds to the situation in which the observations have an upper bound. There is loss of efficiency relative to the MLE; it becomes more pronounced as ξ becomes negative.

2.5 Zempléni's Test of Hypothesis for the GEV Distribution

For testing the hypothesis that a cdf $F \in$ GEV against the general alternative of F being an arbitrary continuous cdf, Zempléni (1991) proposed the test statistic

$$\phi_n(\mathbf{X}) = \sqrt{n} \min_{a,b} \max_{x} |F_n(x) - F_n^2(ax+b)| , \tag{2.26}$$

where

$$F_n(x) = \frac{1}{n} \sum_{i=1}^{n} I_{[\infty,x]}(X_i), \qquad \mathbf{X} = (X_1, \ldots, X_n) \text{ is the sample}.$$

The motivation is that the GEV distribution fulfills the so-called max-stability property, i.e. for any integer m there exist a_m, b_m such that

$$F(x) = F^m(a_m x + b_m) \quad \text{for all } x \in R .$$

To evaluate the maximum in (2.26), it is sufficient to consider the points X_i and X_i^- $(i = 1, \ldots, n)$. Zempléni (1991) provided algorithms to optimize in b for a fixed a, and the second step is to find the optimum value of a. Both algorithms have the same features: first the optimal value is approximated by an iterative procedure and then the exact solution is achieved via simple calculations. Zempléni also shows that:

$$\Pr(\phi_n(\mathbf{X}) > r) \leq \Pr\left(\sqrt{n} \min_n \max_{x \in \{X_1, \ldots, X_n\}} |U_n(x) - U_n^2(x^a)| > r\right),$$

where U_n denotes the empirical d.f. of the uniform sample. Note that $U(x) = U^2(\sqrt{x})$ for the uniform distribution U; thus the value of a converges to 0.5 as $n \to \infty$. Based on this result, Zempléni constructed (by simulation) critical values of the ϕ_n statistic (Table 2.2).

The Zempléni test is conservative. In the regular case $(\xi \geq -1)$, the rejection of the hypothesis is less frequent than desired, but as the scale parameter increases the estimates become more accurate. The distribution of ϕ_n depends on the shape parameter, implying that

Table 2.2: The quantiles of the estimator $\phi_n(\cdot)$.

	$n = 50$	$n = 100$	$n = 200$	$n = 500$	$n = 1000$
0.1	0.583	0.604	0.616	0.630	0.635
0.2	0.636	0.664	0.676	0.693	0.699
0.5	0.775	0.800	0.808	0.826	0.834
0.8	0.919	0.956	0.976	0.988	0.997
0.9	1.015	1.044	1.065	1.088	1.094
0.95	1.078	1.125	1.147	1.174	1.177
0.99	1.226	1.282	1.303	1.338	1.350
0.999	1.43	1.46	1.49	1.53	1.54

the test is not similar with respect to ξ. It is also quite unlikely that ϕ_n has a distribution-free limit.

The power of the test — as the calculations by Zempléni show — is adequate. The results indicate a very strong dependence on the shape of the distribution. Distributions with a shape similar to the GEV distribution need about five times more observations for a high probability of the correct decision. But even then $n = 500$ is a large enough sample to render a correct decision at the level $\alpha = 0.05$ in only 50% of the samples. Fortunately, in environmental applications samples of size 500 and larger are very often available. The test is not consistent; there exists a distribution F which is not GEV such that $F(x) = F^2(ax+b)$ for suitably selected a and b. Evidently, additional studies are required.

The test also needs an iterative estimation procedure, which can however be completed on a PC for sample sizes of some hundreds in a few seconds. For larger samples faster computers (or more effective languages) are helpful. By using a simple FORTRAN algorithm on a SUN workstation, Zempléni was able in early nineties to analyze samples up to 5000 elements. An interesting application of this test is discussed in Sec. 2.8.

2.6 Estimation of Tail Index of a Distribution

Techniques for drawing inferences about the tail behavior of a distribution are by now well developed, and most of them are based on the extreme value limit distributions or related families. These methods could roughly be divided into two types: procedures worked out by Hill (1975), Pickands (1975), Weissman (1978), and others based on extreme order statistics and procedures advocated by Smith and Weissman (1985), R. L. Smith (1987), and Davison and Smith (1990) based on observations above a high threshold utilizing the generalized Pareto distribution.

The distribution to be discussed is:

$$G_\xi(x) = \exp - (1 + \xi x)^{-1/\xi},$$

where ξ is a real parameter and x is such that $1+\xi x > 0$. (For $\xi = 0$, we interpret $(1+\xi x)^{-1/\xi}$ as e^{-x}.) Compare with (2.21) in Sec. 2.3.

As stated above, this distribution is one of the extreme value distributions such that for some constants $a_n > 0$ and b_n and some $\xi \in (-\infty, \infty)$:

$$\lim_{n\to\infty} P\left\{\frac{\max\{X_1, X_2, \ldots, X_n\} - b_n}{a_n} \le x\right\} = G_\xi(x) \tag{2.27}$$

for $x \in R$.

The parameter ξ (in the case when $\xi > 0$) is often referred to as the *tail-index* of a distribution and is usually denoted by the letter γ.[a] Much attention has been paid to its

[a] In this text the Greek letter γ is used in several contexts:

(a) γ denotes the Euler's constant:

$$\gamma = \sum_{k=1}^{\infty}\left(\frac{1}{k} - \ln\frac{k+1}{k}\right) = \int_0^1 \left(\frac{1}{1+x} - \frac{1}{\ln x}\right) = 0.57726\,.$$

(b) γ denotes a particular quantile ($0 < \gamma < 1$).
(c) γ denotes (occasionally) the shape parameter of an extreme value distribution.
(d) γ denotes the tail index of a generalized extreme values distribution (almost identical with c).
(e) γ denotes parameters in a linear regression.

estimation. It is based on the pioneering works of Pickands (1975) and Hill (1975) which appeared almost simultaneously in the same journal. Let $X_{(1,n)} \leq X_{(2,n)} \leq \cdots \leq X_{(n,n)}$ be ascending order statistics of $X_1, \ldots, X_n$. For $\gamma \in R$ and $1 \leq k \leq [n/4]$, Pickands (1975) proposed the estimate

$$\hat{\gamma}_n^{(p)} = (\log 2)^{-1} \log \frac{X_{(n-k,n)} - X_{(n-2k,n)}}{X_{(n-2k,n)} - X_{(n-4k,n)}} \tag{2.28}$$

and proved its weak consistency. However, the Pickands estimator is very sensitive to the choice of the intermediate order statistics which are used for estimation: even a small alteration of k can yield a considerable change in the estimate. Dekker and de Haan (1989) provided a natural and general conditions under which $\sqrt{k}(\hat{\gamma}_n^{(p)} - \gamma)$ is asymptotically normal. Among these conditions are $k = k(n) \to \infty$ and $k/n \to 0$ as $n \to \infty$.

For positive γ we have the Hill (1975) estimator

$$H_n = M_n^{(1)} = \frac{1}{k} \sum_{i=0}^{k-1} \log X_{(n-i,n)} - \log X_{(n-k,n)}, \tag{2.29}$$

which uses $k+1$ upper order statistics; note that Pickands estimator uses $X_{(n-k,n)}$, $X_{(n-2k,n)}$ and $X_{(n-4k,n)}$ only. Mason (1982) showed that $M_n^{(1)}$ is a weak consistent estimator provided the sequence $k = k(n) \to \infty$ and $k(n)/n \to 0$ as $n \to \infty$. Deheuvels *et al.* (1988) showed that $M_n^{(1)}$ is a strong consistent estimate provided $k(n)$ is such that $k/\log\log n \to \infty$ and $k(n)/n \to 0$ as $n \to \infty$. They have proved asymptotical normality of $\sqrt{k}(M_n^{(1)} - \gamma)$ with mean 0 and variance γ^2. Asymptotic normality of the Hill estimator was also studied by Beirlant and Teugels (1989). For asymptotic normality to be valid some additional conditions are needed. See, e.g., the comprehensive paper by Dekkers *et al.* (1989) and R. L. Smith (1987) to be discussed below. Cheng and Pan (1995) showed that under certain assumptions on the underlying distribution:

$$P(H_n \leq x) = \Phi(x) + \frac{(1-x^2)\phi(x)}{3k_n^{1/2}} + o\left(\frac{1}{k_n^{1/2}}\right)$$

holds uniformly on $x \in R$ for k_n satisfying

$$k_n \to \infty \qquad \text{and} \qquad k_n = O(n^{\varepsilon}),$$

where $\varepsilon \in (0, 1)$.

(Here $\Phi(x)$ and $\phi(x)$ are the c.d.f. and the density of the standard normal distribution respectively.)

Marohn (1997) proved that the Hill estimator is asymptotically efficient in the sense of Fisher and Wolfowitz and is asymptotically minimax. He also discusses joint estimation of the scale parameter with the tail index.

Berred (1992) constructed from record values two estimators of γ in case $\gamma > 0$. Define, as usual, sequences of record times and record values, $\tau(n)$ and $X(n)$, by

$$\tau(1) = 1, \qquad \tau(n+1) = \min\{j : X_j > X_{\tau(n)}\}, \quad n \geq 1$$

and

$$X(n) = X_{\tau(n)}, \qquad n \geq 1.$$

Berred's (1992) estimators can then be written as

$$R_{k,n}^1 = \frac{1}{k}(\log X(n) - \log X(n-k)),$$

$$R_{k,n}^2 = \frac{1}{nk - k(k-1)/2} \sum_{j=1}^{k} \log X(n-i+1),$$

where the integers $k = k(n)$ involved in $R_{k,n}^1$ satisfy $k(n) \to \infty$ and $\frac{k(n)}{n} \to 0$ as $n \to \infty$ and in $R_{k,n}^2$, $1 \le k < n$ is fixed. Berred (1992) proved that both $R_{k,n}^1$ and $R_{k,n}^2$ are consistent and under very mild conditions are asymptotically normal. Qi (1998) extended Berred's result for $\gamma \in R$.

As it was stated earlier, the basic idea of Pickands is that the conditional distribution function of $X-\mu$ given $X > \mu$ can be approximated by a generalized Pareto distribution (GPD) and the shape parameter of this GPD is an estimator of γ.

Let $X_1, \ldots, X_n$ be i.i.d. random variables with continuous cdf F and let $X_{1n} \ge \cdots \ge X_{nn}$ be their order statistics. Consider k upper extremes $X_{1n}, \ldots, X_{kn}$. Let $x_0 = \{x : F(x) < 1\}$ be the upper end point of F.

Denote

$$F_u(y) = \frac{F(u+y) - F(u)}{1 - F(u)},$$

$0 < y < x_0 - u$, $u < x_0$ (the conditional distribution function of $X - u$ given $X > u$).

Let $G(y; \sigma, \gamma)$ be the generalized Pareto cdf $G(y; \sigma, \gamma) = 1 - (1 - \gamma y/\sigma)^{1/\gamma}$. A basic result due to Pickands (1975) is:

$$\lim_{u \to x_0} \sup_{0<y<x_0-u} |F_u(y) - G(y; \sigma, \gamma)| = 0\,.$$

This fact was utilized by R. L. Smith (1987) in proving the asymptotic normality of estimators of γ for the case $\gamma > -1/2$.

Other estimators of γ are given by the so-called *kernel* estimators (Csörgo *et al.* (1985), Beirlant *et al.* (1996)):

$$\widehat{CDM} = \frac{\sum_{j=1}^{k} (j/k) K(j/k) [\log(X_{n-j+1}^*) - \log(X_{n-j}^*)]}{\int_0^1 K(t)dt}, \tag{2.30}$$

where K denotes a non-negative nonincreasing kernel defined on (0, 1). The Hill estimator is obtained from $\widehat{CDM}$ by taking $K(u) = \mathbf{1}_{(0,1)}(u)$. Csörgö *et al.* (1985) derived the asymptotic normality and the bias of the kernel estimators. In finding efficient estimators, the adaptive choice of the number k of extreme order statistics used in the estimation procedure is of specific interest. (This problem is similar to the choice of a bandwidth in nonparametric density estimation methods.) In extreme value problems bias disappears for small values of k. Due to high volatility of the estimators under consideration, the choices of the number of order statistics to use in estimation is, however, difficult.

Beirlant *et al.* (1996) showed that tail index estimators can be considered as estimates of the slope at the right upper tail of a Pareto quantile plot using weighted least squares algorithms. They suggested algorithms for searching the order statistic to the right of which one obtains an optimal linear fit of the quantile plot. The weighted least squares estimation method for the slope of the Pareto quantile plot leads back to the class of kernel estimators $\widehat{CDM}$.

The estimators mentioned above all have one common property. When the number of upper order statistics used in estimating γ is small, the variance of the estimator is large. On the other hand, the use of large number of upper order statistics introduces bias in estimation. The object is to balance bias and variance to arrive at an optimal choice of k. The basic tool is the function

$$U(x) = \inf\{y | F(y)^{-1} \geq x\}.$$

De Hann (1984) has shown that for $\gamma > 0$, $F \in D(G_\gamma)$ (i.e. the limiting relation (2.27) is may be, valid for F) if

$$\lim_{t\to\infty} \frac{U(tx)}{U(t)} = x^\gamma$$

(for all $x > 0$, namely U is *regularly varying* with index γ). This is a first-order regular variation condition.

Dekkers and de Haan (1993) and de Haan and Stadtmüller (1992) introduced second order regular variation conditions, which are by far more complicated. All these conditions imply that $F \in D(G_\gamma)$ for an appropriate γ and are explicitly given in Dekkers and de Haan (1989). Dekkers *et al.* (1989) concentrated on estimation of γ for general $\gamma \in R$. Their estimate is

$$\hat{\gamma}_n^{(M)} = M_n^{(1)} + 1 - \frac{1}{2} \frac{M_n^{(2)} - 2(M_n^{(1)})^2}{M_n^{(2)} - (M_n^{(1)})^2},$$

where $M_n^{(1)}$ is the Hill estimator (2.29) and

$$M_n^{(2)} = \frac{1}{k} \sum_{i=0}^{k-1} \{\log X_{(n-i,n)} - \log X_{(n-k,n)}\}^2,$$

provided $x^*(F) = \sup\{x | F(x) < 1\} > 0$. (This can always be achieved by a simple shift.) This estimator is known as the *moment estimator*.

With the second-order regularly varying tail conditions one could determine for $\gamma > 0$ a $k_0 = k_0(n)$ such that for the estimator $\hat{\gamma}_n^{(M)}$ the asymptotic second moment of $\hat{\gamma}_n^{(M)} - \gamma$ is minimal and the corresponding estimator satisfies

$$\sqrt{k_0}(\hat{\gamma}_{n,0}^{(M)} - \gamma) \xrightarrow{d} N(b, 1 + \gamma^2)$$

where b denotes the asymptotic bias. [Dekkers and de Haan (1989).] An analogous result is valid for the cases $\gamma < 0$ and $\gamma = 0$.

For generalized extreme value distribution

$$G_\gamma(x) = \exp - (1 + \gamma x)^{-1/\gamma},$$

we have

$$U(t) = \frac{1}{\gamma}[-\log(1 - t^{-1})]^{-\gamma} - 1, \qquad \gamma \neq 0,$$

and for $\gamma = 0$: $U(t) = -\log(-\log(1 - t^{-1})) = \log t - (2t)^{-1} + o(t^{-1})\ t \to \infty$.

In this case the optimal value $k_0(n)$, $n \to \infty$, is

$$k_o(n) = \begin{cases} \left[\dfrac{(1+\gamma)^4(1+\gamma^2)}{2\gamma^5}\right]^{(1+2\gamma)^{-1}} & n^{2\gamma(1+2\gamma)}(1+o(1)), \quad 0 < \gamma < 1 \\ [8(1+\gamma^2)(2\gamma-1)^{-2}]^{1/3} & n^{2/3}(1+o(1)), \quad \gamma > 1 \\ [64/9]^{1/3} & n^{2/3}(1+o(1)), \quad \gamma = 0. \end{cases}$$

There are many equivalent forms of second-order regular variation; for $\gamma > 0$ one of the forms is that the function $\log U(t) - \gamma \log t - \log c$ or equivalently $t^{-\gamma}U(t) - c$ is regularly varying with index $-\gamma\rho$ for some $\rho > 0$ and $c > 0$.

(Check that for the GVE distribution (with $\gamma > 0$)

$$\log(t^{-\gamma}U(t)/c) = -\gamma t^{-\gamma/2} - t^{\gamma} + o(t^{-2} + t^{-2\gamma}), \qquad t \to \infty$$

so that the above condition is fulfilled for $c = 1/\gamma$ and $\rho = \min(1, \gamma^{-1})$.)

Comparing Pickands (1975) and Dekker *et al.* (1989) estimators, it becomes clear, upon graphing the estimators for varying k, that the moment estimator behaves in a much more stable way and conclusions can be drawn more easily. The moment estimator is based on an average whereas the Pickands estimator uses only a few order statistics. Also, the moment estimator uses the extreme order statistics whereas the Pickands estimator does not.

An averaged Pickards estimator:

$$\sqrt{k}\frac{1}{k}\left\{\frac{\sum_{i=k}^{2k-1} X_{n-i,n} - X_{n-2k,n}}{X_{n-k,n} - X_{n-2k,n}}\right\}$$

has been suggested. However this estimator has asymptotic variance of the same order as that of Pickands estimator and larger than the Dekkers *et al.* moment estimator and its behavior with k is as unstable as that of Pickands'.

Drees (1995) constructed a mixture of Pickands estimators and introduced a bias correction term which results in an estimator robust against an unsuitable choice of the fraction of largest order statistics used in its formation.

2.7 Other Forms of Generalized Extreme Value Distributions

Some forms of *generalized* and *compound* type 1 extreme value distributions have been constructed by Dubey (1969). He generalizes the distribution by introducing an extra parameter τ, defining the cdf by the equation

$$\Pr[X \le x] = \exp\left[-\tau\sigma \exp\left\{-\frac{x-\mu}{\sigma}\right\}\right]. \tag{2.31}$$

However, since

$$\tau\sigma \exp\left\{-\frac{x-\mu}{\sigma}\right\} = \exp\left\{-\frac{x-\mu'}{\sigma}\right\}$$

with $\mu' = \mu + \sigma \log \tau\sigma$, it can be seen that X still has an ordinary type 1 distribution discussed in Chap. 1. This generalized distribution is, however, introduced only as an intermediate step in the construction of a *compound* (convoluted) type 1 extreme value distribution, which formally can be denoted as:

"Generalized" type 1 extreme value $(\mu, \sigma, \tau) \hat{\tau}$ Gamma (p, β).

Here τ is supposed to have the pdf

$$p_\tau(t) = \frac{\beta^p}{\Gamma(p)} t^{p-1} e^{-\beta t}, \qquad t > 0;\ p > 0,\ \beta > 0\,.$$

The resulting compound distribution has the cdf

$$\begin{aligned}\Pr[X \le x] &= \left[\frac{\beta^p}{\Gamma(p)}\right] \int_0^\infty t^{p-1} \exp\left[-t\left\{\beta + \sigma \exp\left(-\frac{x-\mu}{\sigma}\right)\right\}\right] dt \\ &= \left[1 + \sigma\beta^{-1} \exp\left\{-\frac{x-\mu}{\sigma}\right\}\right]^{-p}.\end{aligned}$$

This distribution can be regarded as a generalized logistic distribution originally cited by Hald (1952). In fact it is often termed type 1 generalized logistic distribution. By considering the cdf

$$\Pr[X \le x] = 1 - \exp\left[-\tau\sigma \exp\left\{\frac{x-\mu}{\sigma}\right\}\right] \tag{2.32}$$

and using a similar gamma compounding, Balakrishnan and Leung (1988a) derived the cdf

$$\Pr[X \le x] = 1 - e^{-p(x-\mu)/\sigma} \left[\sigma\beta^{-1} + \exp\left\{-\frac{x-\mu}{\sigma}\right\}\right]^{-p}. \tag{2.33}$$

This distribution has been termed a type 2 generalized logistic distribution. The type 1 and type 2 generalized logistic distributions are related by a simple negation of the random variables. Balakrishnan and Leung (1988a) also considered the exponential-gamma density function

$$p_X(x|\tau) = \exp\left[-\tau \exp\left\{-\frac{x-\mu}{\sigma}\right\}\right] \exp\left\{-\frac{\kappa(x-\mu)}{\sigma}\right\} \frac{\tau^k}{\sigma\Gamma(\kappa)},$$
$$-\infty < x < \infty, \kappa > 0, \sigma > 0\,, \tag{2.34}$$

and compounded it with a gamma density function for τ to derive the density function

$$\begin{aligned}p_X(x) &= \int_0^\infty e^{-te^{-(x-\mu/\sigma)}} e^{-\kappa(x-\mu)} \frac{t^k}{\sigma\Gamma(k)} \frac{\beta^p}{\Gamma(p)} t^{p-1} e^{-\beta t} dt \\ &= \frac{1}{\sigma B(\kappa, p)} \frac{[\beta^{-1} \exp\{-(x-\mu)/\sigma\}]^\kappa}{[1 + \beta^{-1} \exp\{-(x-\mu)/\sigma\}]^{\kappa+p}},\end{aligned} \tag{2.35}$$

$-\infty < x < \infty$, $\kappa > 0$, $p > 0$, $\sigma > 0$.

The density function in (2.35) has been termed a type 4 generalized logistic density. For the special case when $p = \kappa$, the type 4 generalized logistic density function in (2.35) becomes symmetric about $x = \mu$ and has been referred to as a type 3 generalized logistic density. (There is indeed some confusion in the terminology!)

A two-component mixture of extreme value distributions with the density function

$$p_X(x) = \frac{\alpha}{\sigma} e^{-(x-\mu)/\sigma} e^{-e^{-(x-\mu)/\sigma}} + \frac{1-\alpha}{\sigma^*} e^{-(x-\mu^*)/\sigma^*} e^{-e^{-(x-\mu^*)/\sigma^*}},$$
$$-\infty < x < \infty, \quad 0 < \alpha < 1, \quad \sigma > 0, \quad \sigma^* > 0, \tag{2.36}$$

and the cdf

$$F_X(x) = \alpha e^{-e^{-(x-\mu)/\sigma}} + (1-\alpha) e^{-e^{-(x-\mu^*)/\sigma^*}}, \qquad -\infty < x < \infty, \tag{2.37}$$

has also been used in some applied problems. The moment-generating function of this distribution is

$$M_X(t) = \alpha e^{t\mu}\Gamma(1-\sigma t) + (1-\alpha)e^{t\mu^*}\Gamma(1-\sigma^* t), \qquad |t| \max(\sigma, \sigma^*) < 1. \tag{2.38}$$

In particular, the mean and the variance are

$$E[X] = \{\alpha(\mu-\mu^*) + \mu^*\} + \gamma\{\alpha(\sigma-\sigma^*) + \sigma^*\} \tag{2.39}$$

and

$$\mathrm{var}(X) = \frac{\pi^2}{6}\{\alpha\sigma^2 + (1-\alpha)\sigma^{*2}\} + \alpha(1-\alpha)\{(\mu-\mu^*) + \gamma(\sigma-\sigma^*)\}^2. \tag{2.40}$$

Rossi *et al.* (1986) have made use of this two-component extreme value distribution for flood frequency analysis; also see Beran *et al.* (1986) for additional comments.

Revfeim (1984a) introduced the following extension of the type 1 (Gumbel) extreme value distribution:

Let the events occur in a Poisson process at a rate ρ. If the sizes of the events are distributed independently of their occurrences and of each other (with distribution function $F(x)$), then the maximum sizes within unit time interval have asymptotically the distribution function

$$G(x) = (\exp\{-\rho[1-F(x)]\} - e^{-\rho})/(1-e^{-\rho}).$$

For large ρ (say larger than 5), $e^{-\rho}$ is negligible and if $F(x) = 1 - e^{-x/\mu}$ (an exponential distribution) then

$$G(x) \approx \exp\{-\rho e^{-x/\mu}\} \tag{2.41}$$

(an alternative parametric form of the Gumbel distribution usually written as $\exp\{-e^{-\alpha(x-\mu)}\}$). The cumulants of distribution are

$$\begin{aligned}
\kappa_1 &= \mu(\ln\rho + \gamma) \\
\kappa_2 &= \zeta_2\mu^2 \\
\kappa_3 &= 2\zeta_3\mu^3 \simeq 2.404\mu^3 \\
\kappa_4 &= -\pi^2\mu^4/60 \quad \text{(a negative quantity)},
\end{aligned}$$

here γ is the Euler constant (0.5772) and $\zeta_n = \sum_{k=1}^{\infty} k^{-n}$ are the Riemann zeta functions. Revfeim (1984b) generalized the distribution (2.41) by assuming

$$F_i(x) = 1 - e^{-x/\mu}\sum_{k=0}^{i-1}\left(\frac{x}{\mu}\right)^k \Big/ k!$$

(a family of Gamma distributions).[3] In that case

$$G_i(x) = \exp\left\{-\rho e^{-x/\mu}\sum_{k=0}^{i-1}\left(\frac{x}{\mu}\right)^k \Big/ k!\right\}. \tag{2.42}$$

The raw moments of (2.42) are

$$E(X^j) = \mu^j(\rho/\Gamma(i))\int_0^1 (-\ln y)^{i+j-1}\exp\left\{-\rho y\sum_{k=0}^{i-1}(-\ln y)^k/k!\right\}dy \tag{2.43}$$

with $y = e^{-x/\mu}$.

The integral is somewhat complicated for large i and j having a singularity at $y = 0$. The mean value s shows a near log-linear form for all i and can be approximated by $a\mu(\ln\rho + b)$ with $a = 1.00$, $b = 0.58$ for $i = 1$ and $a = 1.13$, $b = 1.82$ for $i = 2$.

Maximum likelihood estimators of μ and ρ when the shape parameter i is known are given by

$$\hat{\mu} = \frac{\overline{\overline{X}}}{i}\left(1 + \frac{\hat{S}_1}{\hat{S}_0}\right) \quad \text{and} \quad \hat{\rho} = \frac{1}{\hat{S}_0},$$

where

$$S_0 = \sum_{k=0}^{i-1}\frac{Z_k}{k!}, \qquad S_1 = \frac{Z_i}{i!},$$

with

$$Z_m = \frac{1}{n}\sum_{k=1}^{n} e^{x_k/\mu}\left(\frac{x_k}{\mu}\right)^m.$$

Scarf (1992) proposed a modification of GEV when the location parameter is of the form $\mu = \mu_0 t^\beta$, $\sigma = \sigma_0 t^\beta$ leading to a power law dependence of the mean of the GEV distribution on time. (The observations here are pairs (x_i, t_i), $i = 1, \ldots, k$, where x_i is observed at time t_i independently of x_j at time t_j.) We shall denote this distribution by GEV $(\mu t^\beta, \sigma t^\beta, \xi)$.

The second modification is when only the location parameter is a function of time to be denoted by GEV $(\mu t^\beta, \sigma, \xi)$. For model 1, the log-likelihood is

$$l(\beta, \mu, \sigma, \xi; \mathbf{x}, \mathbf{t}) = -n\log\sigma - \beta\sum\log t_i - (1+\xi)\sum_{i=1}^{n} y_i - \sum_{i=1}^{n} e^{-y_i},$$

where $y_i = -\frac{1}{\xi}\log\{1 + \frac{\xi}{\sigma}(x_i t_i^{-\beta} - \mu)\}$, $\xi x_i t_i^{-\beta} > \mu\xi + \sigma$, $i = 1, \ldots, n$.

The likelihood equations have to be solved iteratively as in the case of the three-parameter GEV distribution. Scarf (1992) described the procedure in detail. He also discussed PWM estimation for his model. The difficulty here is that the power law parameter β cannot be associated with any particular probability weighted moment of the distribution GEV $(\mu t^\beta, \sigma t^\beta, \xi)$ or GEV $(\mu t^\beta, \sigma, \xi)$. He, therefore, suggests finding the PWM estimator of the distribution of the transformed variable $Z = Xt^{-\beta}$ which would allow to obtain estimates of the variance–covariance matrix of (μ, σ, ξ) and also an estimate of the variance of β from regression.

For Scarf's model 2 if the observations can be presented in the form: $x_{ij} : j = 1, \ldots, n_i$ at time t_i for each $i = 1, \ldots, n$ and $n_i > 2$ for at least two distinct values of i, one can then estimate μ, σ, ξ at two distinct points in time and thus estimate β. Unfortunately, the method of PWM estimation does not allow estimation of the full variance-covariance matrix for the four-parameter GEV distribution.

A motivation for Scarf's four-parameter distribution is found in applications to metallic corrosion where x_i is the depth of the largest pit penetration over a standard area a of metal surface exposed to a corrosive environment for time t_i, and n is the number of such areas of size a. The separate areas could either be distinct "coupons" (i.e. the metallic specimens) from a designed experiment, or else a random sample of areas, *at various times*, from regions of the metal surface that are representative of the whole metal surface under inspection. See the next section for additional details.

2.8 Some Selected Applications

The GEV distribution was recommended for flood frequency analysis in the U.K. (United Kingdom) Flood Studies Report (Natural Environment Research Council, 1975), and, since the introduction of the index-flood procedure based on PWM estimation by Wallis (1980) and Greis and Wood (1981), it has gained much interest (see e.g. Hosking *et al.* (1985a); Wallis and Wood (1985); Lettenmaier *et al.* (1987); Hosking and Wallis (1988); Chowdhury *et al.* (1991); and Lu and Stedinger (1992)). It is used to model extremes of natural phenomena such as river lengths, sea levels, stream flows, rainfall and air pollutants to obtain distributions of daily or annual maxima. In reliability context analogous analyses are performed when the interest is in sample minima of strengths and failure times.

During the last decade de Haan and, especially, Tawn and his associates have written numerous pioneering papers on application of the generalized extreme value methodology to environmental sciences. It is beyond our scope to cover — however briefly — all the publications by these prolific writers and other contributors. The brief descriptions below should therefore be viewed as a hopefully representative sample of their valuable contributions ingenously blending theory and practice — the ultimate goal of the modern statistical science. For additional applications, up to late eighties, with emphasis on engineering, the reader is advised to consult Castillo (1988).

de Haan (1990) (cf. also de Haan and Dekkers (1990)) presents the results of estimating parameter ξ in the distribution

$$G_\xi(x) = \begin{cases} \exp\left(-(1+\xi x)^{-1/\xi}\right), & \text{for} \quad 1+\xi x > 0 \\ \exp\left(-e^{-x}\right) & \text{for} \quad \xi = 0 \end{cases}$$

based on 1577 *high*-tide water levels observed at the Dutch station Hoek van Holland during the winters from 1887/1888 until 1984/1985.

Coles amd Tawn (1991) investigated sea *w*ave, wave *p*eriod and *s*urge data for Newlyn, Cornwall, measured in years 1971–1977. The data was reduced to 15-hourly maximum of each component. Zampléni (1991) was interested to find out whether the component variables are GEV using his test described in Sec. 2.6.

Coles and Tawn (1990) developed the spatial annual maxima method. The approach is to model the joint distribution of the annual maxima over sites accounting for dependence between the sites, and to model the changes in each of the parameters of the GEV distribution over sites.

Dixon and Tawn (1992) employed the spatial annual maxima method to study trends in extreme sea-levels for data from 62 U.K. coastal sites. The measurements were the height of the ocean surface relative to the adjacent land. For many sites the series of annual maximum data appears to have non-stationary features such as trends. The trends are found to have two principal components: changes in eustatic mean sea-level and in land level. (Management authorities are usually concerned with observed trends while climatologists and oceanographers with eustatic ones.) The former appears to be homogeneous over the entire open coastline; whereas the latter has significant spatial variation in the form of north-south tilting of the U.K. Finally, there is some evidence to suggest that estuary sites (at rivers mouths) have different trends from sites along the neighboring open coastline.

In a later publication, Dixon and Tawn (1998) compared direct and indirect methods for estimation of extreme sea-levels depending on whether or not knowledge of tide is exploited. The simplest direct method is fitting of the GEV distribution to observations of annual maximum hourly sea-level — referred to as the annual maximum method (AMM). Pugh and Vassie (1980), Tawn and Vassie (1989) and Tawn (1992) developed an indirect method — referred to as the joint probability method (JPM) — by exploiting the decomposition of sea level into tide and surge components. Indirect methods, which exploit spatial dependencies, were also developed (Dixon and Tawn, 1995). Comparing the two methods for three U.K. east coast sites as well as through a simulation study, it is found that direct methods are biased and underestimate return levels for long return periods. On the whole, the JPM performs substantially better for many U.K. sea-level data sites.

The JPM can also be used to model extreme sea currents by exploiting the decomposition of sea current into tide and surge components. Robinson and Tawn (1997) extend the JPM to handle directionality, temporal independence and tidal non-stationarity that are present in sea current extremes. This involves using the multivariate extreme value model due to Coles and Walshaw (1994) — see Sec. 3.4.6. They demonstrate their methodology for sea current data from Inner Dowing Light Tower in the North Sea.

A recent joint report by J. M. Vassie, D. L. Blackman, J. A. Tawn and M. J. Dixon (1999) described spatial extreme value analysis of largest annual event data combined with historical annual maxima over 14 sites in the Humber estuary. For each site the extremal data set consists of years containing one of the following:

(a) the 7-largest independent annual events
(b) the annual maximum.

Certain sites cover over 60 years; some sites fewer than 15 years; 5 cites provide no data on the 7-largest independent annual event; and other sites provide no data on the annual maximum, the distances in miles along the estuary reference axis range from 11.2 up to 60 miles.

Spatial extreme value methods — delicate procedures — exploit the spatial coherence of the extremal still water level process up the estuary by estimating spatial parameters using simultaneously all the data from each site. The methodology proposed by Vassie *et al.* (1999), cited above, assumes (incorrectly) the independence between the sites. Smith (1993) proposes a complex procedure to adjust the standard errors for spatial dependence (a procedure which is widely used in other statistical areas).

Robinson and Tawn (1995) — in a fascinating contribution — utilize the GEV distribution to investigate whether an athletic record falls within the support of the distribution of possible performances, i.e. whether the record is better than the ultimate performance predicted from previous data. Specifically, the performance of the Chinese athlete Wang Junxia in Beijing national championship (September, 1993) who ran a record time of 486.11 sec in women's 3000 m flat track is analyzed (to determine, inter alia, whether her performance was drug enhanced). The data considered were the annual minima for the women's 3000 m race over the years 1972–1992 together with Wang's time in 1983. The simplistic analysis — using annual minima only — of constructing a confidence interval on the ultimate minimum possible time is insufficient to reach conclusions on Wang's time. Robinson and Tawn (1995) therefore incorporate data on other international standard performances over 3000 m and similar distances within each year. Incorporating the five best annual times for 3000 m results in a conclusion that the associated confidence interval on the ultimate time with 90% confidence is (430.1, 493.8) sec and thus Wang's value (486.11 sec) is — strictly speaking — within the interval. Incorporating relative performances in 1500 m events results in a tighter 90% confidence interval of (478.4, 495.0) sec. Incorporating specifically Olympic and World championship years the expected improvement gives an interval of (483.6, 497.0) sec. Finally, including Wang's time with the 1993 1500 m world record time results in a wider shifted interval of (445.3, 483.3) sec with a longer lower tail but leads to a suspicion of presence of outliers! The authors' conclusion is that no legal case (presumably based on the British law) can be made that her time is from a different population as it is within a 90% confidence interval for the ultimate time. (The authors inform us that they have at present (1999) some additional results related to this problem which are awaited with great anticipation.)

Coles and Pan (1996) analysed extreme pollution levels focusing on Milan (Italy) where a highly sophisticated network of recording stations has been constructed to monitor pollutant levels. The data collected over a period of 11 years in Milan is examined. The authors model extremal behavior of the NO_2 process, taking into account temporal dependence and non-stationarity in the series. The models point to some very strong "qualitative" aspects of NO_2; in particular, the increased rate of occurrence of extreme levels of NO_2 in conditions of calm winds.

Scarf and Laycock (1996) reviewed a number of extreme value models which have been applied to corrosion problems. Special attention is paid to behavior of corrosion extremes such as the largest pit, thinnest wall, maximum penetration, etc. The models are demonstrated on data from laboratory experiments as well as data collected in industrial settings. Emphasizing that corrosion data are inherently of an extreme nature, the authors claim that statistical considerations may be the *only means* of determining numerical values for predictions of maximum pit and other corrosion characteristics.

Coles and Powell (1995) considered the utility of a simple Bayesian approach to spatial modelling of extreme wind speeds. As it was already noted, they find Bayesian estimates to be much more stable than maximum likelihood estimates especially for the shape parameter which has the greatest influence on long-term extrapolation. Also Bayesian analysis performs favorably for short data sets.

Revfeim distribution (see Sec. 2.7) was applied by Revfeim and Hessel (1984) to model extreme wind gusts and by Zelenhasic (1970) (who independently derived this distribution in a Technical Report) to river flow exceedances.

In a pioneering paper R. L. Smith (1989) applied ideas of extreme value theory to the study of ozone in Houston, Texas. He points out that ground level ozone is a topic of considerable environmental concern since excessive levels of ozone are an indication of high air pollution. His data were hourly readings of ozone in Houston from April 1973 to December 1986. Houston is a city in which the legal threshold (12 parts per 100 million) is exceeded very frequently and magnitudes of exceedances are of importance. Smith (1989) produces evidence to indicate strong seasonal pattern of extreme ozone levels with a peak in the summer months extending through to October. The author mentions that ozone analyses at other sites in Texas yielded far more clear-cut evidence of a downward trend.

Niu (1997) investigated — using extreme value theory for non-stationary times series — tropospheric ozone data in the Chicago area. Probabilities of monthly maximum ozone concentrations exceeding some specific levels are estimated, and the mean rate of exceedances of daily maximum ozone over the U.S. national standard 120 ppb — mentioned above — is also assessed.

Appendix: Graphs of the three basic types

(1) Gumbel distribution (standard)

CDF:

$$G(x) = e^{-e^{-x}}$$

PDF:

$$f(x) = \exp(-x - \exp(-x))$$

Transformation: $g(x) = -\ln(-\ln(G(x))$:

$$g(x) = x$$

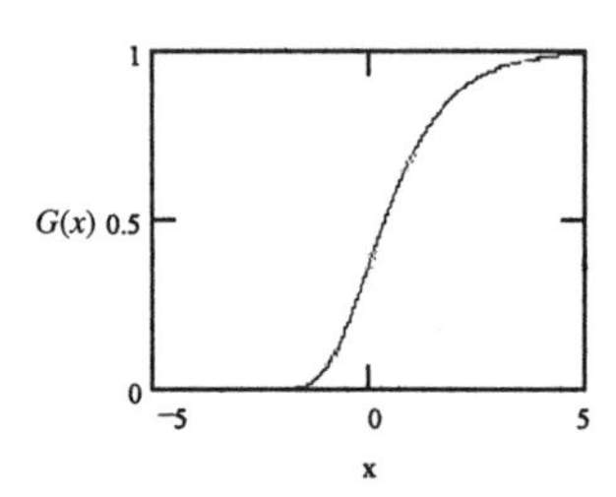

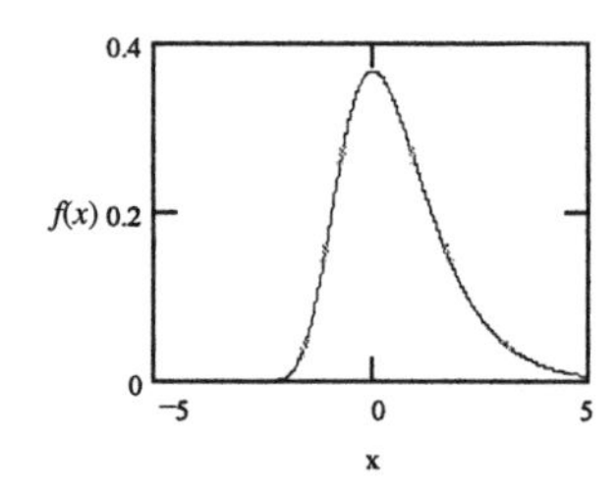

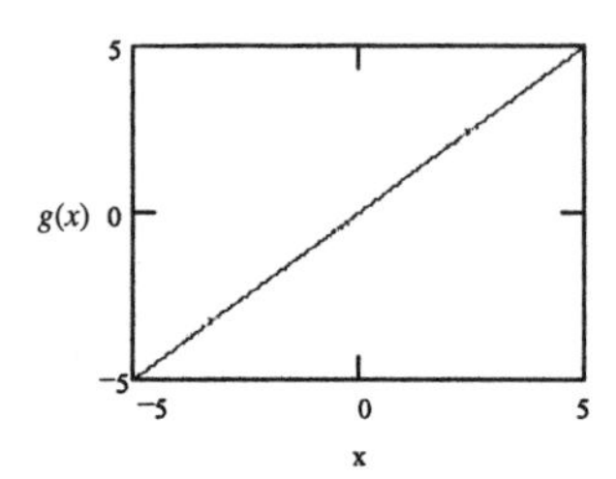

Figure 2.4: Graphs related to the Gumbel distribution.

(2) Fréchet distribution (standard)

CDF:

$$F(x,k) = \begin{cases} e^{-x^{-k}} & x > 0 \\ 0 & x < 0\,. \end{cases} \tag{2.44}$$

PDF:

$$\begin{aligned} f(x,k) &= -x^{(-k)} \cdot \frac{k^2}{x^2} \cdot \exp\left[-x^{(-k)}\right] - x^{(-k)} \cdot \frac{k}{x^2} \exp\left[-x^{(-k)}\right] \\ &\quad + [x^{(-k)}]^2 \cdot \frac{k^2}{x^2} \cdot \exp\left[-x^{(-k)}\right], \qquad x > 0\,. \end{aligned}$$

Transformation: $g(x) = -\ln(-\ln(F(x)))$:

$$g(x,k) = k \ln(x) \qquad \text{(a convex function)}\,.$$

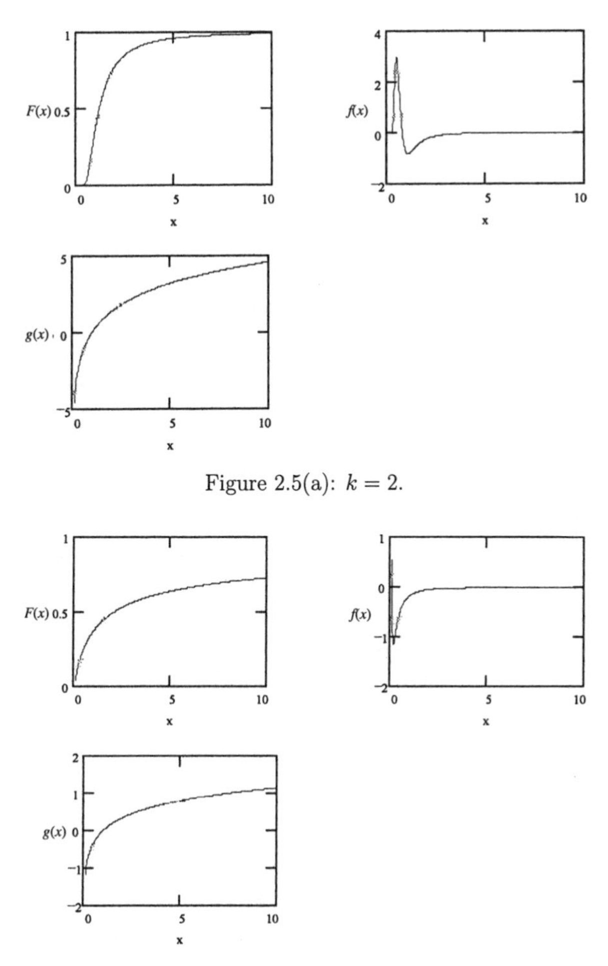

Figure 2.5(a): $k = 2$.

Figure 2.5(b): $k = 0.5$.
Graphs related to the Gumbel distribution.

Weibull distribution (standard)

CDF:

$$W(x,k) = \begin{cases} e^{-(-x)^k} & x < 0 \\ 1 & x > 0 \,. \end{cases} \tag{2.45}$$

PDF:

$$f(x,k) = -(-1)^k x^{(k-1)} k \exp\left[-(-1)^k \cdot x^k\right].$$

Transformation: $g(x) = -\ln(-\ln(W(x)))$:

$$g(x,k) = -k \cdot \ln(-x) \qquad \text{(a concave function)}\,.$$

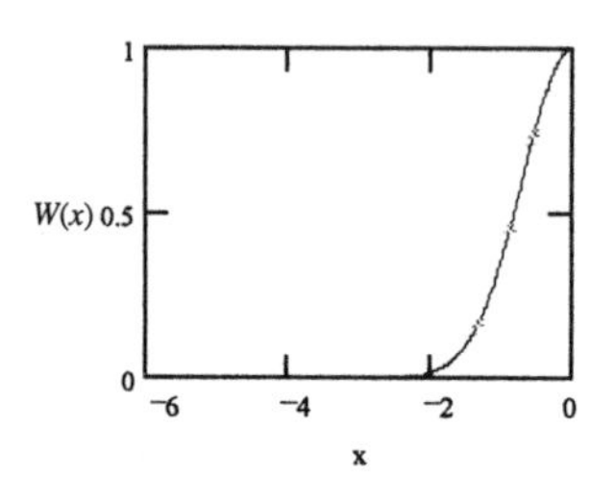

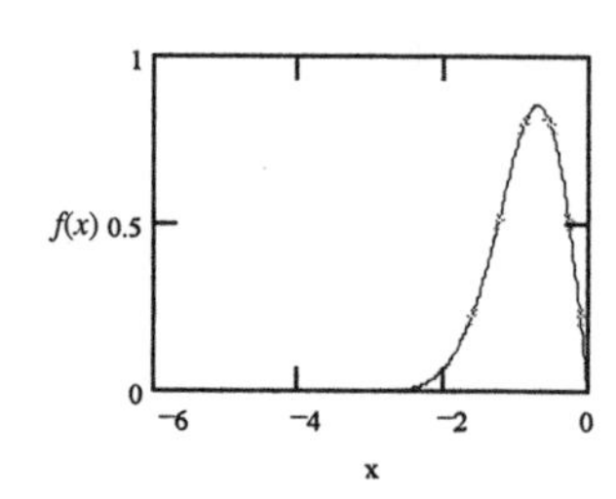

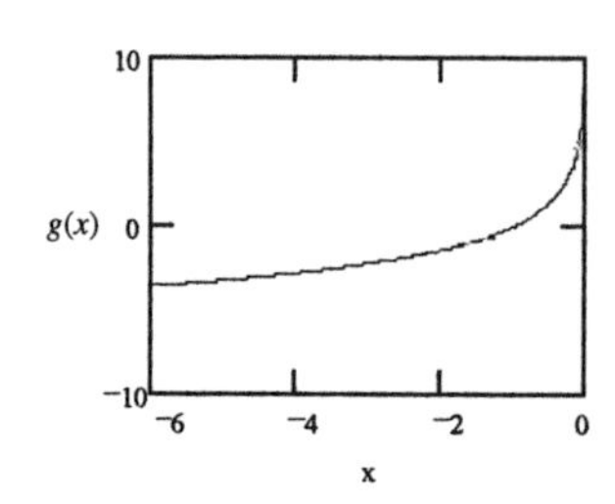

Figure 2.6(a): $k = 2$.

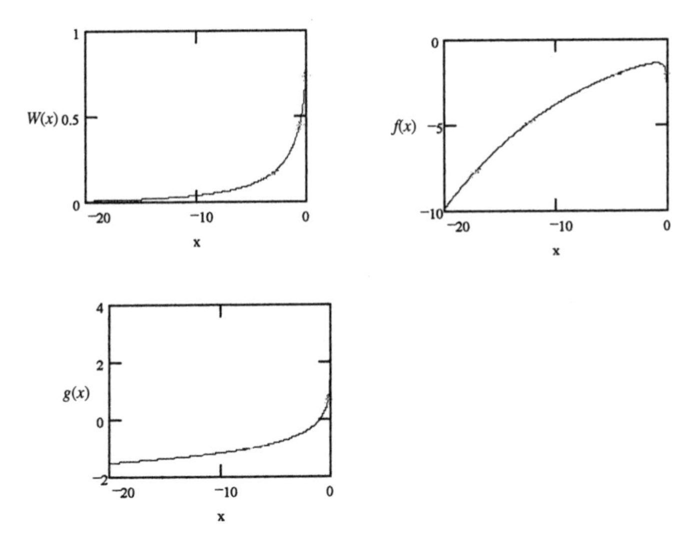

Figure 2.6(b): $k = 0.5$.
Graphs related to the Weibull distribution.

Chapter 3

Multivariate Extreme Value Distributions

The theory of multivariate extreme value distributions is a relatively novel but rapidly growing field. It is somewhat more involved and requires additional concentration. Special effort has been made to provide an organized account of the current state of research. This chapter, consisting of eight sections, presents details of the theory and statistical inference for multivariate extreme value distributions and provides a selective survey of some recent applications.

3.1 Limit Laws for Multivariate Extremes

By analogy with the univariate case the traditional approach to define multivariate extremes is to base it on componentwise maxima. If $\{(X_{i,1},\ldots,X_{i,p}), i=1,\ldots,n\}$ are iid p-variate random vectors with joint df F and

$$\mathbf{M}_n = (M_{n,1},\ldots,M_{n,p}) = \left(\max_{1\le i\le n} X_{i,1},\ldots,\max_{1\le i\le n} X_{i,p}\right)$$

is the vector of maxima of each component, then we seek normalizing constants $a_{n,j}>0$, $b_{n,j}$, $j=1,\ldots,p$ such that as $n\to\infty$

$$\begin{aligned}\Pr\{(M_{n,1}-b_{n,1})/a_{n,1}\le x_1,\ldots,(M_{n,p}-b_{n,p})/a_{n,p}\le x_p\}\\ = F^n(a_{n,1}x_1+b_{n,1},\ldots,a_{n,p}x_p+b_{n,p})\to G(x_1,\ldots,x_p)\end{aligned} \quad (3.1)$$

for a p-variate distribution G with nondegenerate marginals. If this holds for suitable choices of a_n and b_n, then we say G is a multivariate extreme value distribution and F is in the domain of attraction of G, written as $F\in D(G)$. By setting all x_j but one to ∞ in (3.1) we see that $F_j\in D(G_j)$, $j=1,\ldots,p$, i.e.

$$F_j^n(a_{n,j}x_j+b_{n,j})\to G_j(x_j),\qquad j=1,\ldots,p \quad (3.2)$$

where F_j and G_j are the jth marginal distributions of F and G respectively. It follows by the Extremal Types Theorem in Chap. 1 that G_j is a type I, II or III distribution and hence the norming constants $a_{n,j}$, $b_{n,j}$ are precisely those in (1.15)–(1.17).

The two extreme forms of the limiting multivariate distribution correspond to the case of asymptotic total independence between the componentwise maxima for which

$$G(x_1, \ldots, x_p) = G_1(x_1) \cdots G_p(x_p) \tag{3.3}$$

and the case of asymptotic total dependence between the componentwise maxima for which

$$G(x_1, \ldots, x_p) = \min\{G_1(x_1), \ldots, G_p(x_p)\}.$$

Asymptotic total independence arises if and only if (3.2) holds and there exists an $\mathbf{x} = (x_1, \ldots, x_p) \in \Re^p$ such that $0 < G_j(x_j) < 1$, $j = 1, \ldots, p$ and

$$F^n(a_{n,1}x_1 + b_{n,1}, \ldots, a_{n,p}x_p + b_{n,p}) \to G_1(x_1) \cdots G_p(x_p)$$

as $n \to \infty$ (Takahashi, 1994b, Theorem 2.1). Moreover, (3.3) holds for any $(x_1, \ldots, x_p) \in \Re^p$ if and only if

$$G(0, \ldots, 0) = G_1(0) \cdots G_p(0) = \exp(-p),$$

provided that G_j are Gumbel-type with $G_j(x_j) = \exp\{-\exp(-x_j)\}$, $j = 1, \ldots, p$; or,

$$G(1, \ldots, 1) = G_1(1) \cdots G_p(1) = \exp(-p),$$

provided that G_j are Fréchet-type with $G_j(x_j) = \exp\{-x_j^{-\alpha_j}\}$, $\alpha_j > 0$, $j = 1, \ldots, p$; or,

$$G(-1, \ldots, -1) = G_1(-1) \cdots G_p(-1) = \exp(-p),$$

provided that G_j are Weibull-type with $G_j(x_j) = \exp\{-(-x_j)^{\alpha_j}\}$, $\alpha_j > 0$, $j = 1, \ldots, p$ (Takahashi, 1987, Theorems 2.2–2.4). Asymptotic total dependence arises if and only if (3.2) holds and there exists an $\mathbf{x} = (x_1, \ldots, x_p) \in \Re^p$ such that $0 < G_1(x_1) = \cdots = G_p(x_p) < 1$ and

$$F^n(a_{n,1}x_1 + b_{n,1}, \ldots, a_{n,p}x_p + b_{n,p}) \to G_1(x_1)$$

as $n \to \infty$ (Takahashi, 1994b, Theorem 3.1).

To isolate dependence aspects from marginal distributional features it is convenient to transform components of both F and G so that they have a standard marginal distribution. For technical convenience we choose the margins to be described by the unit Fréchet distribution with df $\exp\{-y^{-1}\}$, $y > 0$ denoted by $\Phi_1(y)$. This standardization does not pose difficulties, as shown by the following propositions (Resnick, 1987, Proposition 5.10). Throughout we use the notation Y or y to denote random variables that have the unit Fréchet distribution.

Suppose G is a multivariate df with continuous marginals. Transform

$$G_*(y_1, \ldots, y_p) = G((1/(-\log G_1))^{\leftarrow}(y_1), \ldots, (1/(-\log G_p))^{\leftarrow}(y_p)), \quad y_1 \geq 0, \ldots, y_p \geq 0$$

($\leftarrow$ denotes the inverse of the function in parentheses). Then G_* has marginal distributions $G_{*j}(y) = \Phi_1(y)$ and G is a multivariate extreme value distribution if and only if G_* is also. This proposition standardizes the marginal distributions of a multivariate extreme value df to unit Fréchet margin but yet preserves the extreme value property. The following proposition justifies the standardization by showing that $F \in D(G)$ if and only if $F_* \in D(G_*)$.

Define $U_j = -1/\log F_j$ and $Y_j = U_j(X_j)$ for $1 \le j \le p$. Let F_* be the df of $(Y_1, \ldots, Y_p)$ so that

$$F_*(y_1, \ldots, y_p) = F(U_1^{\leftarrow}(y_1), \ldots, U_p^{\leftarrow}(y_p)).$$

If $F \in D(G)$, then $F_* \in D(G_*)$ and

$$\begin{aligned}\Pr\left(\max_{1\le i\le n} U_j(X_{i,j})/n \le y_j, j = 1, \ldots, p\right) &= F_*^n(ny_1, \ldots, ny_p) \\ &\to G_*(y_1, \ldots, y_p)\end{aligned}$$

as $n \to \infty$. Conversely if $F_* \in D(G_*)$, (3.2) holds and G_* has nondegenerate marginals then $F \in D(G)$.

In the next two sections we provide several fundamental results which characterize the domain of attraction condition, $F \in D(G)$ or equivalently $F_* \in D(G_*)$, and the form of the multivariate extreme value df G_*. These results have been crucial as theoretical underpinnings to recent developments of statistical models for multivariate extremes (see Secs. 3.4–3.7) and their practical applications.

3.2 Characterizations of the Domain of Attraction

The concept of domain of attraction received attention in Chap. 1 for the classical univariate extreme value distributions. The concept is less straightforward for the multivariate case as we shall see from the several characterizations of the domain of attraction that follow. The characterizations are classified into those that are just necessary (Sec. 3.2.1), those that are just sufficient (Sec. 3.2.2) and those that are both necessary and sufficient (Sec. 3.2.3).

3.2.1 *Necessary Characterizations*

These characterizations are especially useful for statistical modelling of multivariate extreme values. We begin with the point process characterization due to de Haan (1985).

Suppose $(X_{i,1}, \ldots, X_{i,p})$, $i = 1, \ldots$ are iid p-variate random vectors with common joint df $F \in D(G)$. Define

$$T(y_1, \ldots, y_p) = \left(\sum_{j=1}^p y_j, y_1\Big/\sum_{j=1}^p y_j, \ldots, y_{p-1}\Big/\sum_{j=1}^p y_j\right) \tag{3.4}$$

and let

$$S_p = \left\{(w_1, \ldots, w_{p-1}) : \sum_{j=1}^{p-1} w_j \le 1, w_j \ge 0, j = 1, \ldots, p-1\right\}$$

denote the $(p-1)$-dimensional unit simplex. Then the point process

$$\mathcal{P}_n = \{(U_1(X_{i,1})/n, \ldots, U_p(X_{i,p})/n), i = 1, \ldots, n\} \to \mathcal{P} \tag{3.5}$$

as $n \to \infty$ where $\mathcal{P}$ is a nonhomogeneous Poisson process on $\Re_+^p$ with intensity measure μ_* satisfying

$$\mu_* \circ T^{\leftarrow}(dr, d\mathbf{w}) = r^{-2} dr H_*(d\mathbf{w}), \quad r > 0, \quad \mathbf{w} \in S_p \tag{3.6}$$

($\circ$ denotes composition of the functions on either side of it) and H_* is a non-negative measure on S_p with

$$H_*(S_p) = p \quad \text{and} \quad \int_{S_p} w_j H_*(d\mathbf{w}) = 1, \quad j = 1, \dots, p-1. \tag{3.7}$$

An immediate consequence is that we can write

$$G_*(y_1, \dots, y_p) = \exp\{-V(y_1, \dots, y_p)\}, \tag{3.8}$$

where

$$\begin{aligned} V(y_1, \dots, y_p) &= \mu_*(([0, y_1] \times \cdots \times [0, y_p])^c) \\ &= \int_{S_p} \max\left(\frac{w_1}{y_1}, \dots, \frac{1 - \sum_{j=1}^{p-1} w_j}{y_p}\right) H_*(d\mathbf{w}). \end{aligned}$$

We refer to V as the exponent measure function.

The intuitive content of (3.5) for $p = 2$ can be described as follows. As $n \to \infty$, the scaling by $1/n$ drags down to the origin all points except those with unusually large values of either $X_{i,1}$ or $X_{i,2}$ or both. Points with unusually large $X_{i,1}$ but not large $X_{i,2}$, will move under the scaling to the horizontal boundary of $\Re_+^2$, and those with unusually large $X_{i,2}$ but not large $X_{i,1}$ will move to the vertical boundary of $\Re_+^2$: only points with both components unusually large will survive in $\Re_+^2$ away from the boundaries.

The limiting intensity measure μ_* describes the dependence structure between unusually large values of $X_{i,j}$, $j = 1, \dots, p$ after standardization by U_j to have the unit Fréchet distribution. However, under the transformation T, which maps the standardized vector $(U_1(X_{i,1}), \dots, U_p(X_{i,p})) \in \Re_+^p$ into pseudo-polar coordinates in $(0, \infty) \times S_p$, the measure μ_* factorizes into a known function of the radial component, r, and a measure H_* of the angular component, $\mathbf{w}$. Thus, essentially, the measure H_* on S_p embodies the dependence structure of the extremes. If it concentrates its mass in the interior of S_p, then we have strong dependence structures, e.g. total dependence between the extremes of $X_{i,j}$, $j = 1, \dots, p$ corresponds to H_* having all its mass at $\{(1/p, \dots, 1/p)\}$, i.e. $H_*(\{(1/p, \dots, 1/p)\}) = p$. If it concentrates its mass near the boundary of S_p, then we have weak dependence structures, e.g., total independence between the extremes corresponds to H_* having all its mass at the vertices, i.e. $H_*(\{(1/p, \dots, 0)\}) = \cdots = H_*(\{(0, \dots, 1/p)\}) = 1$.

Although an arbitrary finite non-negative measure, the standardization of $X_{i,j}$ constrains H_* to have unit means with respect to each dimension of S_p. Since these are the only constraints on H_*, no finite parametrization exists for the measure.

We now discuss two technical tools for generating parametric models for H_* that will be useful later. This requires some terminology. For a given $(w_1, \dots, w_{p-1}) \in S_p$, define

$$H(w_1, \dots, w_{p-1}) = H_*([0, w_1] \times \cdots \times [0, w_{p-1}]),$$

the measure function associated with H_*, and construct $\mathbf{w}^* = (w_1^*, \dots, w_p^*)$, a p-dimensional vector, by setting $w_j^* = w_j$, $j = 1, \dots, p-1$ and $w_p^* = 1 - \sum_{j=1}^{p-1} w_j$. Decompose the measure function H into a hierarchy of densities $h_{m,c}$ defined on subspaces $S_{m,c} = \{\mathbf{w} \in S_p : w_k^* = 0, k \notin c\}$ where $c = \{j_1, \dots, j_m\}$ is an index variable over the subsets of size m of the set $c_p = \{1, \dots, p\}$. The subspace $S_{m,c}$ is isomorphic to the $(m-1)$-dimensional unit simplex S_m and $h_{m,c}$ is the $(m-1)$-dimensional density of H on the subspace $S_{m,c}$. The density $h_{m,c}$

describes the dependence structure between the extremes of $X_{i,k}$ for $k = j_1, j_2, \ldots, j_m$. When $m = p$ and $c = c_p$, we shall simplify the notation by $h \equiv h_{p,c_p}$.

The first tool relates the exponent measure function, V, to H, by expressing the density $h_{m,c}$ for $c = \{j_1, \ldots, j_m\}$ in terms of derivatives of V (Coles and Tawn, 1991, Theorem 1). Namely,

$$\frac{\partial V}{\partial y_{j_1} \cdots \partial y_{j_m}} = -\left(\sum_{l=1}^{m} y_{j_l}\right)^{-(m+1)} h_{m,c}\left(\frac{y_{j_1}}{\sum y_{j_l}}, \ldots, \frac{y_{j_{m-1}}}{\sum y_{j_l}}\right) \tag{3.9}$$

on $\{\mathbf{y} \in \Re_+^p : y_k = 0, k \notin c\}$ where we assume differentiability of V. The importance of this result is that densities of all orders for the measure function H may be obtained for any closed form multivariate extreme value df. For $p = 2$ the result shows the following: H, a function on the unit interval $[0, 1] = S_2$, decomposes into density $h_{2,\{1,2\}}$ defined in the interior, $(0, 1)$, and "densities" $h_{1,\{1\}}$ and $h_{1,\{2\}}$ defined respectively at the end points, $\{1\}$ and $\{0\}$. The two latter "densities" (these are actually atoms of mass, $H_*(\{1\})$ and $H_*(\{0\})$) are independent components of H in that they are associated with those $(X_{i,1}, X_{i,2})$ which are extreme in only one component. The density $h_{2,\{1,2\}}$ is the dependence component in that it describes the dependence between the extremes of both components.

The second tool generates a form for H_* by transforming an arbitrary density $h_\dagger$ in the interior of S_p into h_{p,c_p} (Coles and Tawn, 1991, Theorem 2). Specifically, if $h_\dagger$ is an arbitrary density in the interior of S_p with positive first moments:

$$m_j = \int_{S_p} w_j^* h_\dagger(w_1, \ldots, w_{p-1}) dw_1 \cdots dw_{p-1}, \quad j = 1, \ldots, p, \tag{3.10}$$

then a measure H_* on S_p defined by

$$h_{m,c} \equiv 0, \quad \forall\, c \neq c_p,$$
$$h_{p,c_p}(w_1, \ldots, w_{p-1}) = \frac{1}{m_0} \prod_{j=1}^{p} \frac{m_j}{m_0} h_\dagger\left(\frac{m_1 w_1}{m_0}, \ldots, \frac{m_{p-1} w_{p-1}}{m_0}\right), \tag{3.11}$$

where $m_0 = \sum_{j=1}^{p} m_j w_j^*$, is a valid measure satisfying the constraints (3.7). Hence this result is useful in generating a rich class of parametric models for H_* in the interior of S_p.

The characterization, (3.5), assumes max-stable dependence between the extremes of $(X_{i,1}, \ldots, X_{i,p})$. A generalization of this characterization to cover weaker forms of dependence structures including total independence and negative association is described below (Ledford and Tawn, 1997, Theorem 1). We provide the result for $p = 2$ to the best of our knowledge. As yet it is not known how it generalizes to the multivariate case.

Let $(Y_{i,1}, Y_{i,2})$, $i = 1, \ldots$ be independent random vectors with both Y_1 and Y_2 having the unit Fréchet distribution. Suppose that for y_1 and y_2 simultaneously large

$$\Pr(Y_1 > y_1, Y_2 > y_2) = \mathcal{L}_1(y_1, y_2) y_1^{-c_1} y_2^{-c_2} + \mathcal{L}_2(y_1, y_2) y_1^{-(c_1+d_1)} y_2^{-(c_2+d_2)} + \ldots, \tag{3.12}$$

where $c_1 + c_2 = 1/\eta$, $0 < \eta \leq 1$, $d_k \geq 0$ and $\mathcal{L}_k(y_1, y_2) \not\equiv 0$ denotes a bivariate slowly varying function. Suppose also that $\mathcal{L}_2(ty_1, ty_2) = o\{\mathcal{L}_1(ty_1, ty_2)\}$ as $t \to \infty$ if $d_1 = d_2 = 0$ and

$$g_*(w) = \lim_{t \to \infty} \left\{ \frac{\mathcal{L}_1(tw, t(1-w))}{\mathcal{L}_1(t, t)} \right\}$$

is differentiable for all $w \in (0,1)$. Choose b_n to satisfy

$$\Pr\{b_n^{-1}\max(\min(Y_{1,1},Y_{1,2}),\ldots,\min(Y_{n,1},Y_{n,2})) \le y\} \to \exp\left(-y^{-1/\eta}\right)$$

as $n \to \infty$ and define $T(y_1,y_2) = ((y_1+y_2)/b_n, y_1/(y_1+y_2))$. Then

$$\mathcal{P}_n = \{(Y_{i,1}/b_n, Y_{i,2}/b_n), i = 1,\ldots,n\} \to \mathcal{P} \tag{3.13}$$

as $n \to \infty$ where $\mathcal{P}$ is a nonhomogeneous Poisson process on $(0,\infty)\times(0,\infty)$ with intensity measure μ_* satisfying

$$\mu_* \circ T^{\leftarrow}(dr,dw) = r^{-(1+\eta)/\eta} dr \lambda_0(w) dw, \quad r > 0, w \in (0,1),$$

where $\lambda_0(w)$ is a rather formidable function:

$$\lambda_0(w) = \frac{c_1c_2g_*(w) + w(1-w)g_*'(w)(2w-1+c_1-c_2) - g_*''(w)w^2(1-w)^2}{w^{1+c_1}(1-w)^{1+c_2}}.$$

An immediate consequence is that we can generalize the form of (3.8) by

$$\lim_{n\to\infty} \Pr\left(\max_{1\le i\le n} Y_{i,1} < b_n y_1, \max_{1\le i\le n} Y_{i,2} < b_n y_2\right)$$

$$= \exp\left[-\eta \int_0^1 \lambda_0(w)\left\{\max\left(\frac{w}{y_1}, \frac{1-w}{y_2}\right)\right\}^{1/\eta} dw\right],$$

where the integration is over the open interval $0 < w < 1$.

As for the point process characterization the intensity measure μ_* factorizes into radial and angular components. But here both terms influence the dependence structure with η playing a fundamental role: the $r^{-(1+\eta)/\eta}$ term describes the main decay of probability due to dependence while the $\lambda_0(w)$ term embodies less important features of the dependence. If the common df of $(Y_{i,1}, Y_{i,2})$ belongs to the domain of attraction of G_*, it is then easily verified that $c_1 = c_2 = 1/2$, $d_1 = d_2 = 1/2$, $\eta = 1$ and

$$g_*(w) = \frac{1 - V\{(1-w)^{-1}, w^{-1}\}}{\{2 - V(1,1)\}\{w(1-w)\}^{1/2}}.$$

Thus, $b_n = n$ and (3.13) reduces to (3.5).

Ledford and Tawn (1996) refer to η as the *coefficient of tail dependence* as it provides a measure of the dependence between the marginal tails of Y_1 and Y_2. For example, if $1/2 < \eta \le 1$ the marginal variables are positively associated; when the marginal variables are independent, then $\eta = 1/2$; if $0 < \eta < 1/2$ the marginal variables are negatively associated. Also if the marginal variables are asymptotically dependent then $\eta = 1$, and if $\eta < 1$ then there is asymptotic independence.

Peng (1999) proposes the following consistent estimator for η:

$$\hat{\eta}_n = \log 2 / \log \frac{\sum_{i=1}^n I\{Y_{i,1} > Y_{n,n-2k,1} \text{ and } Y_{i,2} > Y_{n,n-2k,2}\}}{\sum_{i=1}^n I\{Y_{i,1} > Y_{n,n-k,1} \text{ and } Y_{i,2} > Y_{n,n-k,2}\}},$$

where $Y_{n,1,j} \le \cdots \le Y_{n,n,j}$ denote the order statistics of $Y_{1,j},\ldots,Y_{n,j}$ for $j = 1,2$. Peng also establishes asymptotic normality of this estimate by considering the cases $\eta < 1$ and $\eta = 1$ separately. Assume the following variant of (3.12):

$$\Pr\{Y_1 > -1/\log(1-ty_1), Y_2 > -1/\log(1-ty_2)\} = c(y_1,y_2)t^{1/\eta}[1 + O(t^\beta)]$$

uniformly on $\{(y_1, y_2) : y_1^2 + y_2^2 = 1, y_1 \geq 0, y_2 \geq 0\}$ as $t \to 0$ where $\eta \in (0,1]$, $\beta > 0$ and $c(y_1, y_2) \neq 0$ for some $y_1, y_2 > 0$. Assume also that $c(y_1, y_2)$ has continuous first-order partial derivatives denoted by

$$c_j(y_1, y_2) = \frac{\partial c(y_1, y_2)}{\partial y_j}, \quad j = 1, 2.$$

Then for $\eta < 1$

$$\sqrt{\sum_{i=1}^{n} I\{Y_{i,1} > Y_{n,n-k,1}, Y_{i,2} > Y_{n,n-k,2}\}} \sqrt{(2^{1/\hat{\eta}_n})/(2^{1/\hat{\eta}_n} - 1)}(\log 2)\hat{\eta}_n^{-2}(\hat{\eta}_n - \eta) \to^d N(0,1),$$

where $k = k(n)$ is chosen to satisfy

$$k \to \infty, k/n \to 0, k(n/k)^{1-1/\eta} \to \infty \quad \text{and} \quad k(n/k)^{1-2\beta-1/\eta} \to 0$$

as $n \to \infty$. For $\eta = 1$

$$2(\log 2)c(1,1)\sqrt{k}(\hat{\eta}_n - 1) \to^d N(0, \sigma^2), \tag{3.14}$$

where

$$\begin{aligned}\sigma^2 = {} & 2c(1,1)[1 - 4c_1(1,1) - 4c_2(1,1) + 6c_1(1,1)c_2(1,1)] \\ & + 4c(1,2)c_1(1,1)[1 - c_2(1,1)] \\ & + 4c(2,1)c_2(1,1)[1 - c_1(1,1)] + 2c_1^2(1,1) + 2c_2^2(1,1)\end{aligned}$$

and $k = k(n)$ is chosen to satisfy

$$k \to \infty, \qquad k = o(n^{2\beta/(1+2\beta)})$$

as $n \to \infty$. Obviously the limit in (3.14) involves the unknown quantities $c(y_1, y_2)$, $c_1(1,1)$ and $c_2(1,1)$. Peng gives the following consistent estimates:

$$\hat{c}(y_1, y_2) = \frac{1}{k} \sum_{i=1}^{n} I\{Y_{i,1} > Y_{n,n-[ky_1],1}, Y_{i,2} > Y_{n,n-[ky_2],2}\},$$

$$\begin{aligned}\hat{c}_1(1,1) = k^{1/4} \Bigg\{ & \frac{1}{k} \sum_{i=1}^{n} I\{Y_{i,1} > Y_{n,n-[k(1+k^{-1/4})],1}, Y_{i,2} > Y_{n,n-k,2}\} \\ & - \frac{1}{k} \sum_{i=1}^{n} I\{Y_{i,1} > Y_{n,n-k,1}, Y_{i,2} > Y_{n,n-k,2}\} \Bigg\},\end{aligned}$$

$$\begin{aligned}\hat{c}_2(1,1) = k^{1/4} \Bigg\{ & \frac{1}{k} \sum_{i=1}^{n} I\{Y_{i,1} > Y_{n,n-k,1}, Y_{i,2} > Y_{n,n-[k(1+k^{-1/4})],2}\} \\ & - \frac{1}{k} \sum_{i=1}^{n} I\{Y_{i,1} > Y_{n,n-k,1}, Y_{i,2} > Y_{n,n-k,2}\} \Bigg\}.\end{aligned}$$

The next result provides another generalization of the point process characterization, by considering the case where the marginal variables are linearly ordered (Nadarajah *et al.*, 1998, Theorem 2).

Suppose that the joint df F of (X_1, X_2) belongs to the domain of attraction of G. If $X_1 \leq X_2 \leq mX_1$, $m > 1$ then the limit measure H_* in (3.6), defined on $S_2 = [0,1]$, is concentrated in the subinterval

$$\left[\liminf_{y\to\infty}\left\{\frac{l(y)}{y+l(y)}\right\}, \limsup_{y\to\infty}\left\{\frac{y}{y+g(y)}\right\}\right]$$

with

$$\liminf_{y\to\infty}\left\{\frac{g(y)}{y}\right\} \leq 1 \leq \limsup_{y\to\infty}\left\{\frac{y}{l(y)}\right\},$$

where $g(y) = U_2\{U_1^{\leftarrow}(y)\}$ and $l(y) = U_1\{m^{-1}U_2^{\leftarrow}(y)\}$. Consequently, linear ordering between the marginal variables has the effect of reducing the domain of H_* to $[a,b]$ with $a \leq 1/2$ and $b \geq 1/2$. The construct described below provides a simple way of generating parametric models for H_* that are concentrated on a given subinterval $[a,b]$ of $[0,1]$ (Nadarajah *et al.*, 1998, Theorem 3).

Let $H_*^{\dagger}$ be an absolutely continuous positive measure on $[0,1]$ satisfying the constraints (3.7). Let $h^{\dagger}$ denote the density of $H_*^{\dagger}$. Given a subinterval $[a,b]$ of $[0,1]$ with $a \leq 1/2 \leq b$, define a measure H_* on $[a,b]$ as follows: let H_* have atoms of mass

$$H_*(\{a\}) = \gamma_1,$$
$$H_*(\{b\}) = \gamma_2$$

at a and b, where

$$0 \leq \gamma_1 \leq \frac{2b-1}{b-a},$$
$$0 \leq \gamma_2 \leq \frac{1-2a}{b-a},$$

and let H_* be absolutely continuous in the interior (a,b) with density

$$h(w) = \frac{(b-a)(\alpha\beta)^2}{\{\alpha(w-a)+\beta(b-w)\}^3} h^{\dagger}\left\{\frac{\alpha(w-a)}{\alpha(w-a)+\beta(b-w)}\right\}, \quad w \in (a,b),$$

where $\alpha = 2b-1+\gamma_1(a-b)$ and $\beta = 1-2a+\gamma_2(a-b)$. Then H_* satisfies the constraints (3.7).

3.2.2 *Sufficient Characterizations*

Sufficient characterizations enable one to examine whether a given df F_* belongs to the domain of attraction of a multivariate extreme value df G_* and to identify the form of G_*. We provide three sufficient conditions for $F_* \in D(G_*)$. The last two results, in particular, have wide applicability since knowing the limits of some densities enables one to construct the limiting multivariate extreme value distribution.

The first one is based on canonical series expansion of F_* (Campbell and Tsokos, 1973). Suppose $F_*(y_1, y_2)$ satisfies

$$\int_{-\infty}^{\infty}\int_{-\infty}^{\infty}\left[\frac{dF_*(y_1,y_2)}{d\Phi_1(y_1)d\Phi_1(y_2)}\right]^2 d\Phi_1(y_1)d\Phi_1(y_2) < \infty.$$

Suppose also that F_* admits an expansion of the form

$$dF_*(y_1, y_2) = d\Phi_1(y_1)d\Phi_1(y_2)\left\{1 + \sum_{k=1}^{\infty} \rho_k A_k(y_1)B_k(y_2)\right\},$$

where $\{A_k(y_1)\}$ and $\{B_k(y_2)\}$ are the so-called canonical variables defined on $\Phi_1(y_1)$ and $\Phi_1(y_2)$ respectively, and $\{\rho_k\}$ are the canonical correlations defined by

$$\rho_k = \int_{-\infty}^{\infty}\int_{-\infty}^{\infty} A_k(y_1)B_k(y_2)dF_*(y_1, y_2), \qquad k = 1, 2, \ldots.$$

In general, $A_k(y_1)$ and $B_k(y_2)$ are kth order orthonormal polynomials in y_1 and y_2, respectively. (By convention, $A_0(y_1) = B_0(y_2) = 1$.) Then, if F_* belongs to the domain of attraction of G_* it must be of the form

$$G_*(y_1, y_2) = \Phi_1(y_1)\Phi_1(y_2)e^{V_*(y_1,y_2)},$$

where

$$V_*(y_1, y_2) = \lim_{t\to\infty} t\sum_{k=1}^{\infty} \rho_k E[A_k(Y_1)|Y_1 \le ty_1]E[B_k(Y_2)|Y_2 \le ty_2].$$

The second result uses regular variation of the joint density of F_* (De Haan and Resnick, 1987). Suppose F_* has joint density f which is regularly varying with limit function λ, i.e. for $\mathbf{u} = (u_1, \ldots, u_p) \in \Re_+^p\backslash\{(0, \ldots, 0)\}$,

$$\lim_{t\to\infty} t^{p+1}f(tu_1, \ldots, tu_p) = \lambda(u_1, \ldots, u_p).$$

Evidently λ satisfies $\lambda(t\mathbf{u}) = t^{1-p}\lambda(\mathbf{u})$ for $\mathbf{u} \in \Re_+^p\backslash\{(0, \ldots, 0)\}$. Suppose further that λ is bounded on $B = \{\mathbf{u} \in \Re_+^p : \|\mathbf{u}\| = 1\}$ and that the following uniformity condition holds:

$$\lim_{t\to\infty}\sup_{\mathbf{u}\in B} |t^{p+1}f(tu_1, \ldots, tu_p) - \lambda(u_1, \ldots, u_p)| = 0.$$

Then, for any $\epsilon > 0$,

$$\lim_{t\to\infty}\sup_{\|\mathbf{u}\|>\epsilon} |t^{p+1}f(tu_1, \ldots, tu_p) - \lambda(u_1, \ldots, u_p)| = 0.$$

Also λ is integrable on $[\mathbf{0}, \mathbf{y}]^c$, $\mathbf{y} > \mathbf{0}$ and $F_* \in D(G_*)$ where

$$G_*(y_1, \ldots, y_p) = \exp\left\{-\int_{[\mathbf{0},\mathbf{y}]^c} \lambda(\mathbf{u})d\mathbf{u}\right\}, \qquad \mathbf{y} > \mathbf{0}.$$

The final result supposes absolute continuity of F_* (Yun, 1997). For any $c = \{j_1 < \cdots < j_k\} \subset \{1, \ldots, p\}$ with $k \ge 2$, let $f_{j_k|j_1,\ldots,j_{k-1}}(y_{j_k}|y_{j_1}, \ldots, y_{j_{k-1}})$ denote the conditional density of the j_kth component of F_* given values of the $(j_1, \ldots, j_{k-1})$th components. If, for any $c \subset \{1, \ldots, p\}$ with $k \ge 2$,

$$l_{j_k|j_1,\ldots,j_{k-1}}(u_{j_k}; u_{j_1}, \ldots, u_{j_{k-1}}) = \lim_{t\to\infty} tf_{j_k|j_1,\ldots,j_{k-1}}(tu_{j_k}|tu_{j_1}, \ldots, tu_{j_{k-1}}) < \infty$$

and if, in addition, for every fixed $u_{j_1}, \ldots, u_{j_{k-1}}$, there exists a $t^*(u_{j_1}, \ldots, u_{j_{k-1}}) < \infty$ such that the class

$$\{tf_{j_k|j_1,\ldots,j_{k-1}}(tu_{j_k}|tu_{j_1}, \ldots, tu_{j_{k-1}}) : t^*(u_{j_1}, \ldots, u_{j_{k-1}}) < t < \infty\}$$

of functions of u_{j_k} is locally uniformly integrable over $(0,\infty)$, then $F_* \in D(G_*)$ with

$$G_*(y_1,\ldots,y_p) = \exp\Bigg\{ -\sum_{j=1}^{p} \frac{1}{y_j} - \sum_{c\subset\{1,\ldots,p\}:|c|\geq 2} \times (-1)^{|c|-1} \int_{y_{j_1}}^{\infty} \cdots \int_{y_{j_k}}^{\infty} \beta_k(u_1,\ldots,u_k) du_1 \ldots du_k \Bigg\}$$

where

$$\beta_k(u_1,\ldots,u_k) = \frac{1}{u_1^2}\left(\prod_{j=1}^{k-1} l_{j+1|1,\ldots,j}(u_{j+1};u_1,\ldots,u_j)\right).$$

3.2.3 *Necessary and Sufficient Characterizations*

Here we give three results, all necessary and sufficient for $F_* \in D(G_*)$.

The first result is due to Marshall and Olkin (1983) and expresses G_* as the limit of the conditional distribution of $\mathbf{Y}$ given that at least one component of $\mathbf{Y}$ has exceeded t. Namely, $F_* \in D(G_*)$ if and only if

$$\frac{-\log F_*(ty_1,\ldots,ty_p)}{-\log F_*(t,\ldots,t)} \to \frac{-\log G_*(y_1,\ldots,y_p)}{-\log G_*(1,\ldots,1)} \tag{3.15}$$

as $t\to\infty$ for each $y_j > 0$, $j = 1,\ldots,p$.

The second result appears in Resnick (1987, Proposition 5.17(ii)) and involves the limiting intensity measure μ_*. Namely, $F_* \in D(G_*)$ if and only if

$$t \Pr(t^{-1}\mathbf{Y} \in B) \to \mu_*(B) \tag{3.16}$$

as $t\to\infty$ for all relatively compact B for which the boundary of B has μ_* measure equals to 0.

The third result (Takahashi, 1994a, Propositions 2.1 and 2.2) is in terms of

$$D_{F_*}(u_1,\ldots,u_p) = F_*(\Phi_1^{\leftarrow}(u_1),\ldots,\Phi_1^{\leftarrow}(u_p)), \quad (u_1,\ldots,u_p)\in(0,1)^p$$

and

$$D_{G_*}(u_1,\ldots,u_p) = G_*(\Phi_1^{\leftarrow}(u_1),\ldots,\Phi_1^{\leftarrow}(u_p)), \quad (u_1,\ldots,u_p)\in(0,1)^p$$

which are the copulas of F_* and G_* respectively. It says that $F_* \in D(G_*)$ if and only if

$$\lim_{t\to\infty} t[1 - D_{F_*}(\mathbf{u}^{1/t})] = -\log D_{G_*}(\mathbf{u})$$

for all $\mathbf{u}\in(0,1)^p$; or, equivalently,

$$\lim_{t\uparrow 1} \frac{1 - D_{F_*}(\mathbf{u}^{1-t})}{1-t} = -\log D_{G_*}(\mathbf{u})$$

for all $\mathbf{u}\in(0,1)^p$; or, equivalently,

$$\lim_{t\downarrow 0} \frac{1 - D_{F_*}(\mathbf{u}^{t})}{1 - D_{G_*}(\mathbf{u}^{t})} = 1$$

for all $\mathbf{u} \in (0,1)^p$; or, equivalently,

$$d_{j_1,\ldots,j_k}(\mathbf{u}) = \lim_{n\to\infty} n\bar{F}_*(0,\ldots,n\Phi_1^{\leftarrow}(u_{j_1}),\ldots,n\Phi_1^{\leftarrow}(u_{j_k}),\ldots,0) < \infty$$

for all $1 \le j_1 < \ldots < j_k \le p$ and for all $\mathbf{u} \in (0,1)^p$ where $\bar{F}_*$ is the joint survivor function of F_*. If any of these statements is satisfied, then we can write

$$D_{G_*}(\mathbf{u}) = u_1 \cdots u_p \exp\left\{\sum_{k=2}^{p}(-1)^k \sum_{1\le j_1<\ldots<j_k\le p} d_{j_1,\ldots,j_k}(\mathbf{u})\right\}.$$

3.3 Characterizations of Multivariate Extreme Value Distributions

Some of the characterizations in the above section also provided characterizations on the form of G. In this section we consider some more characterizations on the form of G.

The earliest known characterization is that due to Gumbel (1962e). Let $G_{B_1}, G_{B_2}, \ldots, G_{B_m}$ be known bivariate extreme value distributions with unit Fréchet margins. Then, their geometric mean

$$G_{B_1}^{\beta_1}(y_1,y_2) G_{B_2}^{\beta_2}(y_1,y_2) \cdots G_{B_m}^{1-\beta_1-\beta_2-\cdots-\beta_{m-1}}(y_1,y_2)$$

is again a *bivariate extreme value distribution* with *unit Fréchet margins*.

We can directly generalize the stability postulate in (1.5) to obtain the following. Multivariate extreme value dfs, G, in (3.1) are those dfs for which there exist norming constants $\alpha_{n,j} > 0$, $\beta_{n,j}$, $j = 1,\ldots,p$ such that

$$G^n(x_1,\ldots,x_p) = G(\alpha_{n,1}x_1 + \beta_{n,1},\ldots,\alpha_{n,p}x_p + \beta_{n,p}), n \ge 1. \tag{3.17}$$

By setting all x_j but one to ∞ we see that

$$G_j^n(x_j) = G_j(\alpha_{n,j}x_j + \beta_{n,j}),\ j = 1,\ldots,p.$$

Hence, the norming constants $\alpha_{n,j}$, $\beta_{n,j}$ are precisely those for the stability postulate in (1.5).

A characterization due to Tiago de Oliveira (1958) is

$$G_*(y_1,y_2) = \{\Phi_1(y_1)\Phi_1(y_2)\}^{\nu(\log y_2 - \log y_1)},$$

where ν is the so-called dependence function. Obretenov (1991) shows that ν is related to H_* through

$$\nu\left(\log\frac{y}{1-y}\right) = \int_{[0,1]} \max\{w(1-y),(1-w)y\} H_*(dw).$$

For more than two variables the characterization generalizes to

$$G_*(y_1,\ldots,y_p) = \{\Phi_1(y_1)\ldots\Phi_1(y_p)\}^{\nu(\log y_2 - \log y_1,\ldots,\log y_p - \log y_1)}$$

with

$$\nu\left(\log\frac{y_1}{y_2},\ldots,\log\frac{y_1}{y_p}\right)=\int_{S_p}\max\left(\frac{w_1^*y_1}{\sum_{j=1}^p y_j},\ldots,\frac{w_p^*y_p}{\sum_{j=1}^p y_j}\right)H_*(d\mathbf{w}),$$

where S_p is the $(p-1)$-dimensional unit simplex and w_j^* is as defined in Sec. 3.2.1.

An alternative way of writing the point process characterization in (3.8) is as follows (Pickands, 1981). Considering the case $p=2$, we can write

$$G_*(y_1,y_2)=\exp\left\{-\left(\frac{1}{y_1}+\frac{1}{y_2}\right)A\left(\frac{y_1}{y_1+y_2}\right)\right\}, \tag{3.18}$$

where A is also referred to as a dependence function and is related to H_* through

$$A(w)=\int_{[0,1]}\max\{w(1-q),(1-w)q\}H_*(dq).$$

It can be verified that A has the following properties: $A(0)=A(1)=1$; $-1\le A'(0)\le 0$; $0\le A'(1)\le 1$; $A''(w)\ge 0$ and $\max(w,1-w)\le A(w)\le 1$, $0\le w\le 1$; $A(w)=1$ implies that Y_1 and Y_2 are totally independent; $A(w)=\max(w,1-w)$ implies that Y_1 and Y_2 are totally dependent; A is convex, i.e. $A[\lambda y_1+(1-\lambda)y_2]\le\lambda A(y_1)+(1-\lambda)A(y_2)$; and, if A_k are dependence functions, so is $\sum_{k=1}^m\alpha_kA_k$, where $\alpha_k\ge 0$ and $\sum_{k=1}^m\alpha_k=1$.

The next result is a special case of a spectral representation for max-stable processes (De Haan, 1984). There exist non-negative Lebesgue integrable functions $f_j(s)$, $0\le s\le 1$ satisfying

$$\int_{[0,1]}f_j(s)ds=1,\qquad j=1,\ldots,p$$

such that

$$G_*(y_1,\ldots,y_p)=\exp\left\{-\int_{[0,1]}\max\left(\frac{f_1(s)}{y_1},\ldots,\frac{f_p(s)}{y_p}\right)ds\right\}. \tag{3.19}$$

In the next two sections we demonstrate the characterizations described in Secs. 3.2 and 3.3 to develop flexible parametric families for bivariate and multivariate extreme value distributions.

3.4 Parametric Families for Bivariate Extreme Value Distributions

The nine families discussed in this section represent the bulk of the distributions for modelling bivariate extremes. No doubt that additional models will be discovered.

3.4.1 *Logistic Distributions* (Tawn, 1988b)

The df G_* takes the form

$$G_*(y_1,y_2)=\exp\left[-\frac{1-\psi_1}{y_1}-\frac{1-\psi_2}{y_2}-\left\{\left(\frac{\psi_1}{y_1}\right)^q+\left(\frac{\psi_2}{y_2}\right)^q\right\}^{1/q}\right], \tag{3.20}$$

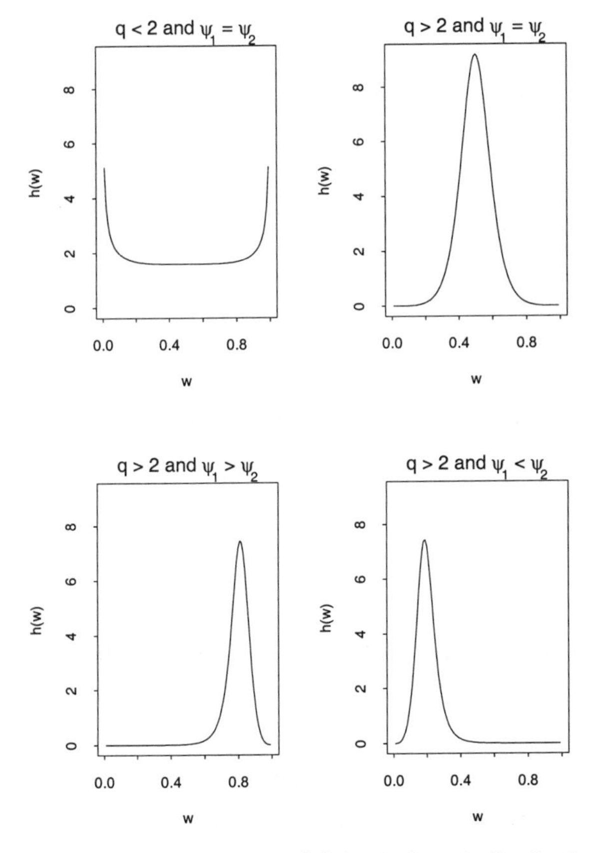

Figure 3.1: Possible forms for $h(w)$ for the logistic distribution.

where $0 \leq \psi_1, \psi_2 \leq 1$ and $q > 1$. Applying (3.9), we have

$$h(w) = (q-1)\psi_1^q\psi_2^q\{w(1-w)\}^{q-2}\{(\psi_2 w)^q + (\psi_1(1-w))^q\}^{1/q-2}$$

and $H_*(\{0\}) = 1 - \psi_2$, $H_*(\{1\}) = 1 - \psi_1$. Thus, this family has mass both in the interior and at the end points. It allows for asymmetry and nonexchangeability through ψ_1 and ψ_2: symmetry and exchangeability arises if and only if $\psi_1 = \psi_2$. Total independence corresponds to $\psi_1 = 0$ or $\psi_2 = 0$ or the limit $q \to 1^+$, whereas total dependence corresponds to $\psi_1 = \psi_2 = 1$ and the limit $q \to \infty$.

A special case for $\psi_1 = \psi_2 = 1$ is the symmetric logistic distribution having all its mass in the interior:

$$G_*(y_1, y_2) = \exp\{-(y_1^{-q} + y_2^{-q})^{1/q}\}. \tag{3.21}$$

This distribution appears in the survival analysis literature; see, for example, Hougaard (1986). Alternative parametrizations for this distribution are possibly advisable: for example $s = 1/q$ $(0 \leq s \leq 1)$. The variables of this distribution are exchangeable and have correlation

$(q^2-1)/q^2$. Also the Fisher information matrix for this distribution has been derived by Oakes and Manatunga (1992).

If $\psi_1 = \psi_2$ we get a mixture of symmetric logistic and independence. If $q \to \infty$, we have

$$G_*(y_1, y_2) = \exp\left\{-\max\left(\frac{1}{y_1} + \frac{1-\psi_2}{y_2}, \frac{1-\psi_1}{y_1} + \frac{1}{y_2}\right)\right\}$$

with $\Pr(Y_1\psi_2 = Y_2\psi_1) = \psi_1\psi_2/(\psi_1 + \psi_2 - \psi_1\psi_2)$. When $\psi_1 = 1$ and $\psi_2 = \alpha$, we have the biextremal (α) distribution:

$$G_*(y_1, y_2) = \exp\left[-\frac{1-\alpha}{y_2} - \left\{\left(\frac{1}{y_1}\right)^q + \left(\frac{\alpha}{y_2}\right)^q\right\}^{1/q}\right],$$

whereas when $\psi_1 = \alpha$ and $\psi_2 = 1$ we have the dual of the biextremal (α) distribution

$$G_*(y_1, y_2) = \exp\left[-\frac{1-\alpha}{y_1} - \left\{\left(\frac{\alpha}{y_1}\right)^q + \left(\frac{1}{y_2}\right)^q\right\}^{1/q}\right],$$

which corresponds to Y_1 and Y_2 being exchanged. If $\psi_1 = \psi_2 = \alpha$ we have the Gumbel distribution (Gumbel and Mustafi, 1967):

$$G_*(y_1, y_2) = \exp\left[-\frac{1-\alpha}{y_1} - \frac{1-\alpha}{y_2} - \alpha\left\{\left(\frac{1}{y_1}\right)^q + \left(\frac{1}{y_2}\right)^q\right\}^{1/q}\right].$$

3.4.2 *Negative Logistic Distributions* (Joe, 1990)

The df G_* takes the form

$$G_*(y_1, y_2) = \exp\left[-\frac{1}{y_1} - \frac{1}{y_2} + \left\{\left(\frac{\psi_1}{y_1}\right)^q + \left(\frac{\psi_2}{y_2}\right)^q\right\}^{1/q}\right],$$

where $0 \le \psi_1,\ \psi_2 \le 1$ and $q < 0$. Applying (3.9), we have

$$h(w) = (1-q)\psi_1^q\psi_2^q\{w(1-w)\}^{q-2}\{(\psi_2 w)^q + (\psi_1(1-w))^q\}^{1/q-2}$$

and $H_*(\{0\}) = 1 - \psi_2$, $H_*(\{1\}) = 1 - \psi_1$. This family is similar in structure to the logistic family with the special case $\psi_1 = \psi_2 = 1$ giving a symmetric version of the family and the limiting cases $q \to 0^-$ and $q \to -\infty$ reducing the family to being totally independent and totally dependent respectively.

3.4.3 *Bilogistic Distributions* (Joe *et al.*, 1992)

This family is motivated by the max-stable representation (3.19). Setting $f_1(s) = (1 - q_1^{-1})s^{-1/q_1}$ and $f_2(s) = (1-q_2^{-1})(1-s)^{-1/q_2}$ into (3.19), we get its df as

$$G_*(y_1, y_2) = \exp\left[-\int_{[0,1]} \max\left\{\frac{(q_1-1)s^{-1/q_1}}{q_1 y_1}, \frac{(q_2-1)(1-s)^{-1/q_2}}{q_2 y_2}\right\} ds\right]$$

for $q_1 > 1$ and $q_2 > 1$. Applying (3.9), we have

$$h(w) = \frac{(1-1/q_1)(1-z)z^{1-1/q_1}}{(1-w)w^2\{(1-z)/q_1 + z/q_2\}},$$

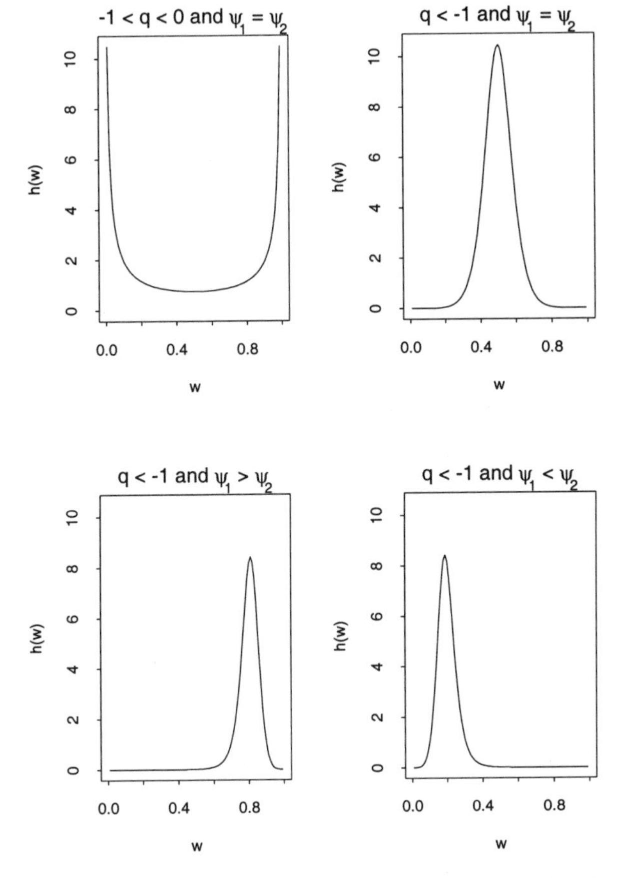

Figure 3.2: Possible forms for $h(w)$ for the negative logistic distribution.

where $z = z(w; q_1, q_2)$ is the root of

$$(1 - 1/q_1)(1 - w)(1 - z)^{1/q_2} - (1 - 1/q_2)wz^{1/q_1} = 0, \tag{3.22}$$

and $H_*(\{0\}) = H_*(\{1\}) = 0$. Thus, this family has all its mass in the interior and is an asymmetric generalization of the logistic family in that setting $q = q_1 = q_2$ gives the symmetric logistic distribution with the two variables being exchangeable. Total independence and total dependence correspond to taking both q_1 and q_2 to 1^+ and ∞ respectively.

It is possible to think of $(q_1 + q_2)/2$ as a dependence parameter, measuring the strength of dependence between the extremes of the two variables, and $(q_1 - q_2)$ as an asymmetry parameter, the case $q_1 - q_2 = 0$ being one in which the two variables are exchangeable. Joe *et al.* (1992) apply this distribution to estimate likely combinations of sulphate and nitrate levels in acid rain.

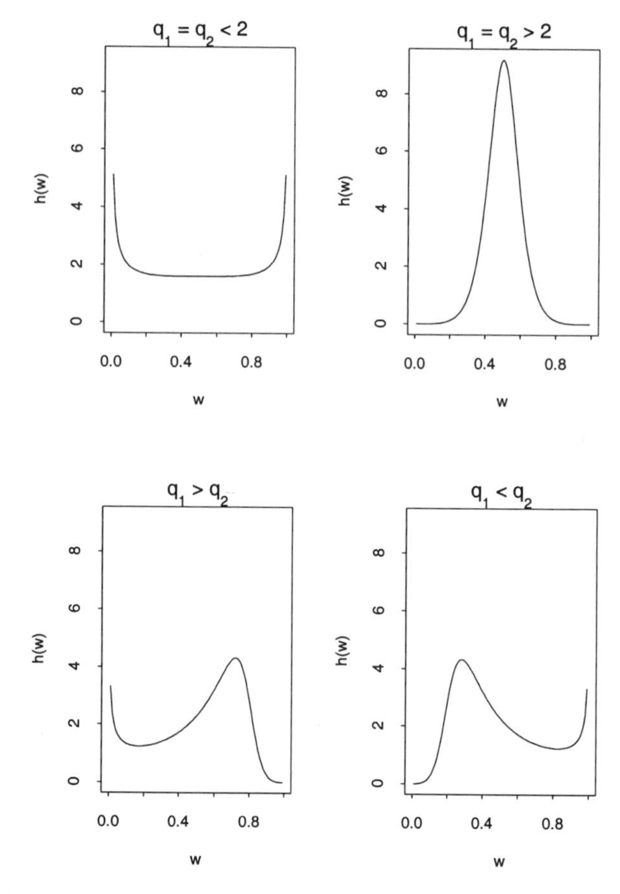

Figure 3.3: Possible forms for $h(w)$ for the bilogistic distribution.

3.4.4 *Negative Bilogistic Distributions* (Coles and Tawn, 1994)

This family has the same df as the bilogistic distributions except that $q_1 < 0$, $q_2 < 0$. Applying (3.9), we have

$$h(w) = -\frac{(1-1/q_1)(1-z)z^{1-1/q_1}}{(1-w)w^2\{(1-z)/q_1 + z/q_2\}}, \quad q_1 < 0, \quad q_2 < 0,$$

and $H_*(\{0\}) = H_*(\{1\}) = 0$ where $z = z(w; q_1, q_2)$ is as defined in (3.22). This family is similar in structure to the bilogistic family and again setting $q = q_1 = q_2$ reduces it to a symmetric and exchangeable version; namely, the symmetric negative logistic family. Now limiting both q_1 and q_2 to 0^- and $-\infty$ correspond to total independence and total dependence respectively. Coles and Tawn (1994) find this distribution most suitable for estimating the dependence between the extremes of surge and wave height.

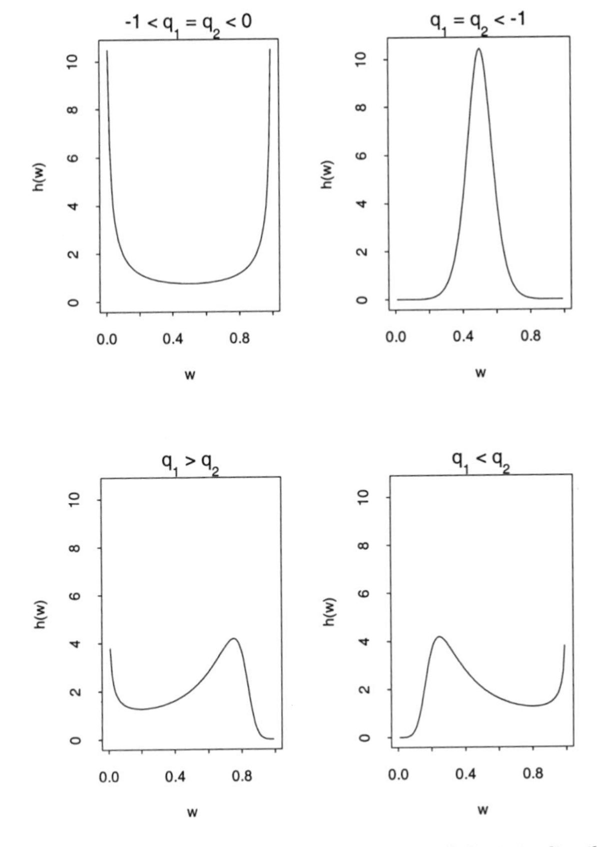

Figure 3.4: Possible forms for $h(w)$ for the negative bilogistic distribution.

3.4.5 *Gaussian Distributions* (Smith, 1991)

The standard Normal distribution is the most prominent distribution in all applications of probabilistic and statistical methodology and it is therefore only natural to find its applications among bivariate extreme value distributions. The joint df has the form:

$$G_*(y_1, y_2) = \exp\left[-\int_{[0,1]} \max\left\{\frac{f_0(s-t_1)}{y_1}, \frac{f_0(s-t_2)}{y_2}\right\} ds\right],$$

where f_0 is the pdf of the Normal(0, σ) distribution. This can be rewritten as

$$G_*(y_1, y_2) = \exp\left[-\frac{1}{y_2}\Phi\left\{s\left(\frac{y_1}{y_1+y_2}\right)\right\} - \frac{1}{y_1}\Phi\left\{a - s\left(\frac{y_1}{y_1+y_2}\right)\right\}\right],$$

where $s(w) = \{a^2 + 2\log w - 2\log(1-w)\}/(2a)$, $a = \{(t_1 - t_2)/\sigma\}^2$ and Φ is the cdf of the standard Normal distribution. Smith (1991) uses this family to model spatial variation of extreme storms at locations corresponding to t_1 and t_2. This family also appears in Husler and Reiss (1989) as the limit distribution of componentwise maxima of independently and identically distributed bivariate Normal vectors; namely, if $\{(X_{i,1}, X_{i,2})\}$ are iid standard Normal random vectors and ρ_n is the correlation coefficient between $X_{i,1}$ and $X_{i,2}$ then

$$\Pr\left\{\max_{1\le i\le n} X_{i,1} \le b_n + x_1/b_n, \max_{1\le i\le n} X_{i,2} \le b_n + x_2/b_n\right\} \to G_*(e^{x_1}, e^{x_2})$$

as $n \to \infty$. Here we suppose that $(1-\rho_n)\log n \to a^2/4$ as $n \to \infty$. The normalizing constant b_n is given by $b_n = n\exp(-b_n^2/2)/\sqrt{2\pi}$. See also Hooghiemstra and Husler (1996) for a similar characterization based on maxima of the projections of iid bivariate Normal vectors with respect to two arbitrary directions. An expression for the measure density h can be derived by applying (3.9) as usual. The resulting form has all its mass in the interior. The value of a controls the amount of dependence with the limits $a \to \infty$ and $a \to 0$ corresponding to total independence and total dependence respectively.

3.4.6 *Circular Distributions* (Coles and Walshaw, 1994)

This family serves as yet another motivation of (3.19). The joint df

$$G_*(y_1, y_2) = \exp\left[-\int_{[0,2\pi]} \max\left\{\frac{f_0(\omega;\theta_1,\zeta)}{y_1}, \frac{f_0(\omega;\theta_2,\zeta)}{y_2}\right\} d\omega\right],$$

where

$$f_0(\omega;\beta,\zeta) = \frac{1}{2\pi I_0(\zeta)}\exp\{\zeta\cos(\omega-\beta)\}$$

is the pdf of the well-known von Mises circular distribution with I_0 denoting the modified Bessel function of order 0. Coles and Walshaw (1994) use this distribution to model the dependence between the extremes of wind speeds corresponding to directions θ_1 and θ_2. Suppose without loss of generality that $\theta_2 \ge \theta_1$, $\theta_2 - \theta_1 \le \pi$ and $\tilde{\theta} = (\theta_2 - \theta_1)/2$. Routine calculations then show that we can rewrite

$$G_*(y_1, y_2) = \exp\left\{-\frac{\int_B f_0(\omega;\tilde{\theta},\zeta)d\omega}{y_1} - \frac{\int_{\tilde{B}} f_0(\omega;-\tilde{\theta},\zeta)d\omega}{y_2}\right\},$$

where

$$B = \{\omega \in (0, 2\pi] : \sin\omega > \gamma(w)\}, \qquad \tilde{B} = (0, 2\pi]\backslash B,$$

and $\gamma(w) = \{\log w - \log(1-w)\}/(2\zeta\sin\tilde{\theta})$. An expression for h can be obtained by straightforward application of (3.9). Like the Gaussian distributions, this family has the mass of h confined to the interior. Here both ζ and angular separation, $\tilde{\theta}$, control the dependence. The strength of dependence increases with ζ while decreasing with $\tilde{\theta}$. The limits $\zeta = 0$ and $\zeta \to \infty$ give total dependence and total independence respectively.

3.4.7 *Beta Distributions* (Coles and Tawn, 1991)

The pdf of Beta(q_1, q_2) distribution is

$$h_\dagger(w) = \frac{\Gamma(q_1+q_2)}{\Gamma(q_1)\Gamma(q_2)} w^{q_1-1}(1-w)^{q_2-1}, \; q_1 > 0, \; q_2 > 0, \; w \in (0,1).$$

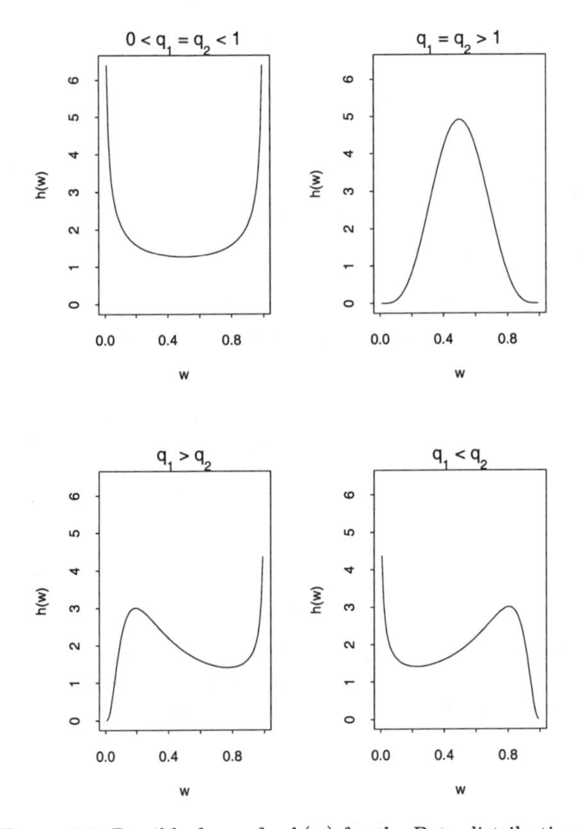

Figure 3.5: Possible forms for $h(w)$ for the Beta distribution.

This distribution is widely used in modelling of hydrological variables (see e.g. Johnson *et al.*, 1995, Vol. 2, p. 236). By eq. (3.10) $m_j = q_j/(q_1+q_2)$, and from eq. (3.11) it follows that

$$h(w) = \frac{q_1^{q_1} q_2^{q_2} \Gamma(q_1+q_2+1)}{\Gamma(q_1)\Gamma(q_2)} \frac{w^{q_1-1}(1-w)^{q_2-1}}{\{q_1 w + q_2(1-w)\}^{1+q_1+q_2}}, \; w \in (0,1)$$

is the density of a valid measure H_* on $[0,1]$ that satisfies the constraints (3.7) with $H_*(\{0\}) = H_*(\{1\}) = 0$. Like the two bilogistic families this is asymmetric, nonexchangeable and has the mass confined to the interior. For the symmetric and exchangeable version (which arises when $q = q_1 = q_2$) both total independence and total dependence are attained as limiting cases by taking $q \to 0^+$ and $q \to \infty$ respectively.

Applying eq. (3.8), we have the corresponding df

$$G_*(y_1, y_2) = \exp\left[-\frac{1}{y_1}\left\{1 - Be\left(q_1+1, q_2; \frac{q_1 y_1}{q_1 y_1 + q_2 y_2}\right)\right\} - \frac{1}{y_2} Be\left(q_1, q_2+1; \frac{q_1 y_1}{q_1 y_1 + q_2 y_2}\right)\right],$$

where

$$Be(\alpha_1, \alpha_2; u) = \frac{\Gamma(\alpha_1 + \alpha_2)}{\Gamma(\alpha_1)\Gamma(\alpha_2)} \int_0^u w^{\alpha_1 - 1}(1 - w)^{\alpha_2 - 1} dw,$$

a normalized incomplete beta function.

3.4.8 *Polynomial Distributions* (Nadarajah, 1999a)

This distribution encompasses the structure of all known bivariate extreme value distributions. A possible motivation is as follows.

One common feature of the distributions in Secs. 3.4.1–3.4.7 is that their structure is governed by the behavior of h near the end points of $[0, 1]$. For example, for the bilogistic distribution we have $h(w) = O(w^{q_2 - 2})$ and $h(1 - w) = O(w^{q_1 - 2})$ as $w \to 0$, and knowing these gives an idea of the whole structure of h and hence that of G_*. In general we can write $h(w) = O(w^r)$ and $h(1 - w) = O(w^s)$ as $w \to 0$ (Nadarajah, 1994). Thus, a natural choice for h that has all the flexibility of the known distributions is:

$$h(w) = \begin{cases} \alpha w^r & \text{if } 0 < w < \theta \\ \beta(1 - w)^s & \text{if } \theta < w < 1 \end{cases}$$

for $\theta \in (0, 1)$ with

$$H_*(\{0\}) = \gamma_0, \qquad H_*(\{1\}) = \gamma_1, \qquad H_*(\{\theta\}) = \gamma_\theta.$$

To ensure non-negativity of h and its continuity at θ we take $\alpha \geq 0$, $\beta \geq 0$ and impose the requirement $\alpha\theta^r = \beta(1 - \theta)^s$. To ensure validity of the unit-mean condition (3.7) we take $r > -1$, $s > -1$ and parametrize the atoms at the end points as:

$$\begin{aligned} \gamma_0 &= 1 - (1 - \theta)\gamma_\theta - \frac{\beta}{s + 2}(1 - \theta)^{s+2} + \alpha\theta^{r+1}\left[\frac{\theta}{r + 2} - \frac{1}{r + 1}\right], \\ \gamma_1 &= 1 - \theta\gamma_\theta - \frac{\alpha}{r + 2}\theta^{r+2} + \beta(1 - \theta)^{s+1}\left[\frac{1 - \theta}{s + 2} - \frac{1}{s + 1}\right] \end{aligned} \tag{3.23}$$

with $0 \leq \gamma_0$, $\gamma_1 \leq 1$ and $0 \leq \gamma_\theta \leq \min\{\theta^{-1}, (1 - \theta)^{-1}\}$. The resulting distribution has, in total, five free parameters. The parameters α and β represent coefficients of the amount of dependence put by h on either side of θ. Large values of them are associated with strong dependencies. The parameters r and s represent the structure of dependence exhibited by h on either side of θ. Negative values of them are associated with weak dependence structures as in that case h puts most of its mass near the end points. The parameter θ represents asymmetry of the dependence structure exhibited by h and also enables accommodation of atoms of mass in the interior. The parameter γ_θ is a measure for the mass of H_* to be concentrated at a single point in the interior (for total dependence the mass of H_* is concentrated at the point $w = 1/2$ with probability one). Finally, γ_0 and γ_1 are measures for the mass of H_* to be concentrated at the end points 0 and 1 respectively (for total independence the mass of H_* is concentrated at each end point with probability half).

It is easily checked that the forms of H and V associated with the distribution are:

$$H(w) = \begin{cases} \gamma_0 + \dfrac{\alpha}{r + 1} w^{r+1} & \text{if } 0 < w < \theta \\ 2 - \gamma_1 - \dfrac{\beta}{s + 1}(1 - w)^{s+1} & \text{if } \theta \leq w < 1 \end{cases}$$

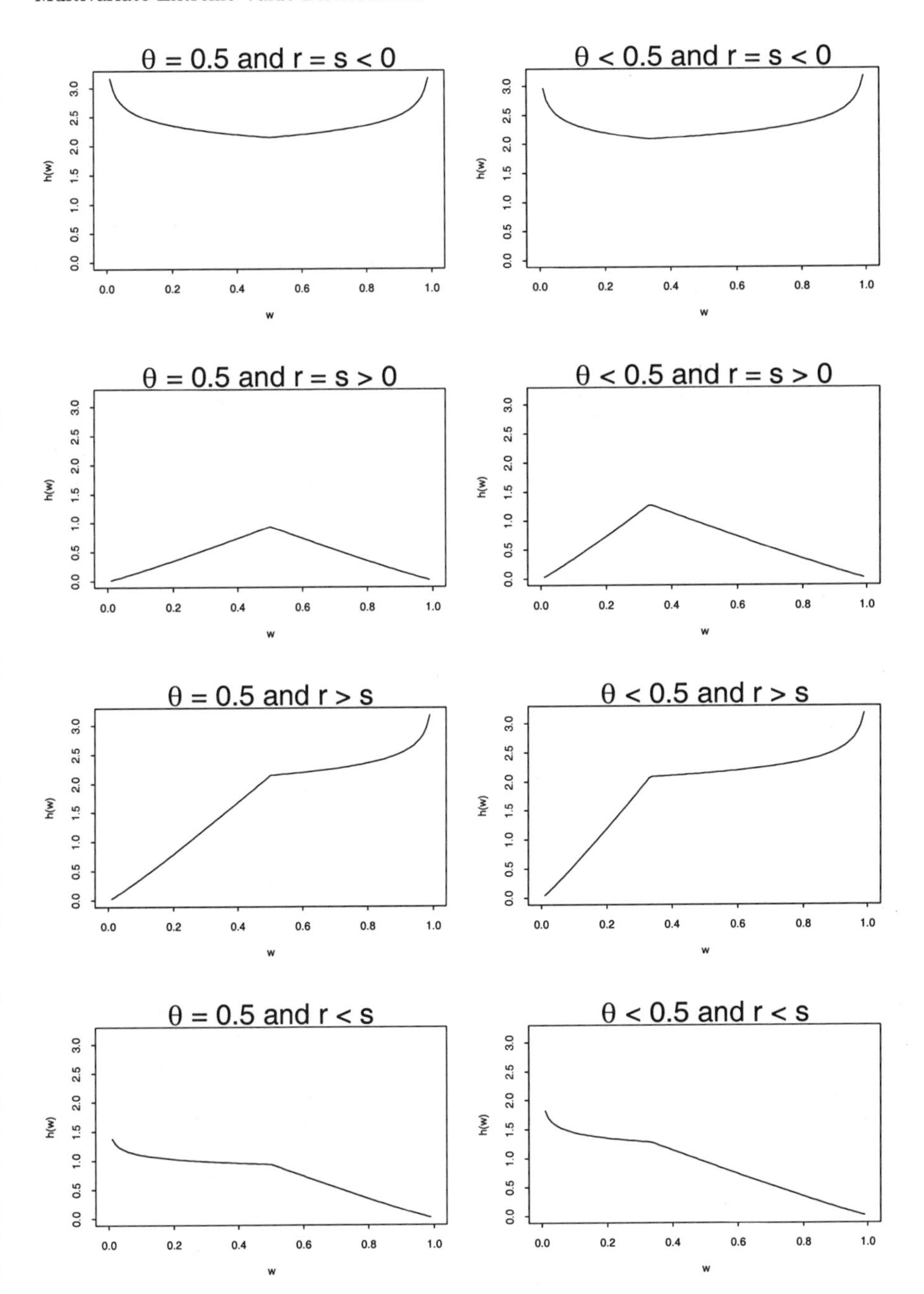

Figure 3.6: Possible forms for $h(w)$ for the polynomial distribution.

and for $y_1 > 0$, $y_2 > 0$

$$V(y_1, y_2) = \begin{cases} \dfrac{1}{y_1} + \dfrac{\gamma_0}{y_2} + \dfrac{1}{y_2}\dfrac{\alpha}{(r+1)(r+2)}\left(\dfrac{y_1}{y_1+y_2}\right)^{r+1} & \text{if } y_1/(y_1+y_2) < \theta \\ \dfrac{1}{y_2} + \dfrac{\gamma_1}{y_1} + \dfrac{1}{y_1}\dfrac{\beta}{(s+1)(s+2)}\left(\dfrac{y_2}{y_1+y_2}\right)^{s+1} & \text{if } y_1/(y_1+y_2) \geq \theta \end{cases}$$

where γ_0, γ_1 are given by (3.23).

The distribution has the requirement that h is continuous at θ, i.e. $\alpha\theta^r = \beta(1-\theta)^s$ and this admits valid solutions for θ for all possible signs of r and s (the solution is unique when r and s have the same signs). Thus, continuity of h at θ is a sensible requirement. However, we find that further requirements for the smoothness of h limit applicability of the distribution. For example, h is differentiable at θ only if $0 \leq \theta = r/(r-s) \leq 1$.

We find that symmetry arises if and only if either $\alpha = \beta$ and $r = s$ when $\theta = 1/2$ or $\alpha = \beta$ and $r = s = 0$ when $\theta \neq 1/2$. Thus, asymmetry of dependence structure for the distribution can be attributed to θ not being equal to $1/2$, the polynomial coefficients not being equal or the polynomial powers not being equal. Exchangeability is equivalent to symmetry when $\theta = 1/2$; otherwise, in addition to symmetry, we must have $\gamma_\theta = 0$. Total independence arises as the special case for $\alpha = 0$, $\beta = 0$ and $\gamma_\theta = 0$ while total dependence arises as the special case for $\alpha = 0$, $\beta = 0$, $\theta = 1/2$ and $\gamma_\theta = 2$.

Two further special cases of interest are $\gamma_0 = \gamma_1 = \gamma_\theta = 0$, where H_* has no atoms of mass, and $\alpha = \beta = 0$, where H_* has no mass in the interiors $(0, \theta)$ and $(\theta, 1)$. In the first case, using conditions (3.23), we can parametrize

$$\alpha = \frac{(r+1)(r+2)\{2(s+1)\theta - s\}}{[(r+2)+(s-r)\theta]\theta^{r+1}}$$

with $\theta \geq s/\{2(s+1)\}$ to ensure $\alpha \geq 0$ and

$$\beta = \frac{(s+1)(s+2)\{(r+2) - 2(r+1)\theta\}}{[(r+2)+(s-r)\theta](1-\theta)^{s+1}}$$

with $\theta \leq (r+2)/\{2(r+1)\}$ to ensure $\beta \geq 0$. Then the continuity requirement on h reduces to the following quadratic equation:

$$\frac{2(r-s)}{(r+2)}\theta^2 + \frac{s^2 - 3rs - 2r}{(r+1)(s+1)}\theta + \frac{s}{s+1} = 0$$

which admits valid solutions for θ for all possible signs of r and s (the solution is unique when r and s have the same signs). The resulting distribution has, in total, two free parameters. In the second case, the mass of H_* is distributed only at the end points and θ. From (3.23) we see that $\gamma_0 = 1 - (1-\theta)\gamma_\theta$ and $\gamma_1 = 1 - \theta\gamma_\theta$. Thus, total independence and total dependence arise when the mass at θ takes the values 0 and 2 (with $\theta = 1/2$) respectively. Exchangeability arises when the mass at both the end points are equal which occurs if and only if $\theta = 1/2$ or the distribution is totally independent. The end point 0 has no mass if and only if $\gamma_\theta = 1/(1-\theta)$ and $\gamma_1 = (1-2\theta)/(1-\theta)$ with $\theta \leq 1/2$. The end point 1 has no mass if and only if $\gamma_\theta = 1/\theta$ and $\gamma_0 = (2\theta - 1)/\theta$ with $\theta \geq 1/2$.

Since, under weak dependence structures H_* concentrates most of its mass near the end points of $[0, 1]$, natural measures of weakness of dependence are:

$$M_1 = \int_0^\theta \left(1 - \frac{w}{\theta}\right) H_*(dw) = 1 - (1-\theta)\gamma_\theta - \frac{\beta}{s+2}(1-\theta)^{s+2} - \frac{\alpha}{r+2}(1-\theta)\theta^{r+1}$$

for the mass in $[0,\theta]$ and

$$M_2 = \int_\theta^1 \left(1 - \frac{1-w}{1-\theta}\right) H_*(dw) = 1 - \theta\gamma_\theta - \frac{\alpha}{r+2}\theta^{r+2} - \frac{\beta}{s+2}\theta(1-\theta)^{s+1}$$

for the mass in $(\theta,1]$. Since, under strong dependence structures H_* concentrates most of its mass in the interior of $[0,1]$, natural measures of strength of dependence are:

$$M_3 = \int_0^\theta \frac{w}{\theta} H_*(dw) = \gamma_\theta + \frac{\alpha}{r+2}\theta^{r+1}$$

for the mass in $[0,\theta]$ and

$$M_4 = \int_\theta^1 \frac{1-w}{1-\theta} H_*(dw) = \frac{\beta}{s+2}(1-\theta)^{s+1}$$

for the mass in $(\theta,1]$. It follows that $M_1+M_2 = 2-\theta^{-1}(1-\theta)^{-1}+V(\theta,1-\theta)$ is the measure of overall weakness of dependence with values of 2 and 0 for total independence and total dependence respectively. Similarly $M_3+M_4 = \theta^{-1}(1-\theta)^{-1} - V(\theta,1-\theta)$ is the measure of overall strength of dependence with values of 0 and 2 for total independence and total dependence respectively. If $\theta = 1/2$, then $(M_1+M_2)/2+1 = V(1,1)$ is the *extremal coefficient* developed by Coles and Tawn (1994) to measure dependence. Clearly larger values of the polynomial coefficients α and β have the effect of strengthening dependence while larger values of the polynomial powers r and s have the reverse effect. Note further that M_1+M_3 and M_2+M_4 are the total mass of H_* in $[0,\theta]$ and $(\theta,1]$ respectively. Clearly the total mass in each segment becomes inflated and deflated respectively with larger values of the polynomial coefficient and power associated with that segment. Note too that $M_1+M_2+M_3+M_4 = 2$, the total mass of H_*.

Some obvious measures of asymmetry are θ, $|r-s|$, $|\alpha-\beta|$, $r/(r+s)$ and $\alpha/(\alpha+\beta)$. Additional measures based on the dependence measures above are $|M_1-M_2|$, $|M_3-M_4-\gamma_\theta|$, $M_1/(M_1+M_2)$ and $(M_3-\gamma_\theta)/(M_3+M_4-\gamma_\theta)$. We have $M_1 = M_2$ if and only if $\theta = 1/2$ or the distribution is totally independent while $M_3 - \gamma_\theta = M_4$ if and only if $\theta = (r+2)/(r+s+4)$.

3.4.9 *Polynomial Distributions* (Kluppelberg and May, 1999)

These are analogous to the above distributions, but formulated in terms of the $A(.)$ function in (3.18). Take

$$A(w) = a_m w^m + a_{m-1}w^{m-1} + \cdots + a_2 w^2 - \left(\sum_{k=2}^m a_k\right) w + 1$$

for $w \in [0,1]$ with $a_2 \geq 0$, $\sum_{k=2}^m a_k \geq 0$, $0 \leq \sum_{k=2}^m (k-1)a_k \leq 1$ and $\sum_{k=2}^m k(k-1)a_k \geq 0$. Then the corresponding joint df

$$G_*(y_1,y_2) = \exp\left\{-\frac{1}{y_1} - \frac{1}{y_2} + \sum_{k=2}^m a_k \sum_{l=0}^{m-k} \binom{m-k}{l} \frac{y_1^{l+k-1} y_2^{m-k-l-1}}{(y_1+y_2)^{m-1}}\right\}$$

has $(m-2)$ parameters. Applying eq. (3.9), we have

$$h(w) = m(m-1)a_m w^{m-2} + (m-1)(m-2)a_{m-1}w^{m-3} + \cdots + 2a_2$$

and $H_*(\{0\}) = H_*(\{1\}) = 0$. Setting $m = 5$, $a_5 = \psi_1/20$, $a_4 = \psi_2/12$, $a_3 = -(\psi_1 + \psi_2)/6$ and $a_2 = 1/2$, we have as a special case

$$h(w) = \psi_1 w^3 + \psi_2 w^2 - (\psi_2 + \psi_1)w + 1,$$

the measure density of the asymmetric mixed distribution due to Tawn (1988b).

3.5 Parametric Families for Multivariate Extreme Value Distributions

The five specific models of multivariate extreme value distributions discussed here do not of course exhaust all possible configurations. Section 3.5.6 provides tools for constructing further multivariate extreme value distributions subject to constraints on their marginals.

3.5.1 *Logistic Distributions* (Tawn, 1990)

The logistic families (Secs. 3.4.1 and 3.4.2) have direct generalizations to the multivariate case. They have been among the most applied multivariate extreme value distribution in the literature. One possible way to motivate the generalization is as follows.

Let C be an index variable over the set B, the class of all nonempty subsets of $\{1, \ldots, p\}$. Let $Z_{j,C}^{(i)}$ be the size of the ith realization at site j, of the extreme spatial storms of the type which occur only at the collection C of sites. Here $Z_{j,C}^{(i)}(i = 1, \ldots, N_C)$ are assumed to be conditionally independent given N_C, where the random variable N_C is taken to have a Poisson distribution of mean τ_C. Also α_C denotes the unrecorded covariate information variable, which has a positive stable distribution and characteristic exponent $0 < 1/q_c \leq 1$. The α_C are assumed to be independent.

We say that a storm affects site j only if an observation at site j exceeds a high threshold, t_j, during that storm. Hence $Z_{j,C}^{(i)} > t_j$ for all $j \in C$ and $Z_{j,C}^{(i)} \leq t_j$ for all $j \notin C$. As discussed in Chap. 2, exceedances of a high threshold have the Generalized Pareto distribution, so for all i we take

$$\Pr(Z_{j,C}^{(i)} < z | Z_{j,C}^{(i)} > t_j) = 1 - \{1 + \xi_j(z - t_j)/\sigma_j\}^{-1/\xi_j},$$

where $z > t_j$, $1 + \xi_j(z - t_j)/\sigma_j > 0$, $\sigma_j > 0$ and $\xi_j \in \Re$. If

$$Z_{j,C} = \max(Z_{j,C}^{(1)}, \ldots, Z_{j,C}^{(N_C)})$$

for $N_C > 0$ then it follows that, for $z > t_j$,

$$\begin{aligned}
\Pr\{Z_{j,C} < z\} &= \sum_{n=1}^{\infty} \left\{ \prod_{i=1}^{n} \Pr(Z_{j,C}^{(i)} < z | Z_{j,C}^{(i)} > t_j) \right\} \Pr(N_C = n) + \Pr(N_C = 0) \\
&= \sum_{n=0}^{\infty} [1 - \{1 + \xi_j(z - t_j)/\sigma_j\}^{-1/\xi_j}]^n \tau_C^n \exp(-\tau_C)/n! \\
&= \exp[-\tau_C\{1 + \xi_j(z - t_j)/\sigma_j\}^{-1/\xi_j}]. \qquad (3.24)
\end{aligned}$$

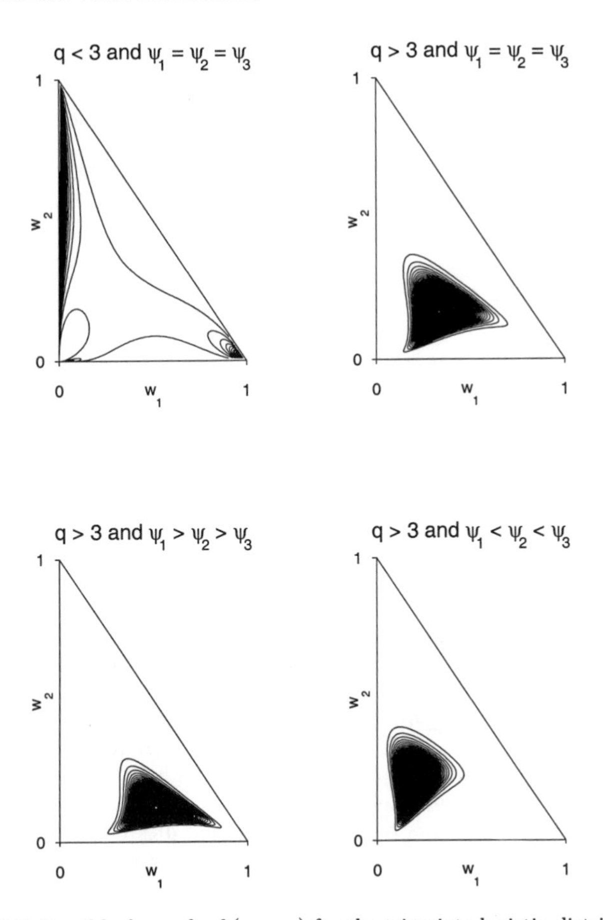

Figure 3.7: Possible forms for $h(w_1, w_2)$ for the trivariate logistic distribution.

As we are interested only in large z, that is $z > t_j$, there is no loss of generality in taking (3.24) to apply for all z such that $1 + \xi_j(z - t_j)/\sigma_j > 0$. Hence $Z_{j,C}$ has a Generalized Extreme Value distribution (see Chap. 2).

Here interest is in the joint behavior of $X_1, \ldots, X_p$ where for $j = 1, \ldots, p$

$$X_j = \max_{C \in B_{(j)}} (Z_{j,C}), \tag{3.25}$$

where $B_{(j)}$ is the subclass of B containing all nonempty subsets which include j. Thus, here maximization is over all spatial storms of the type that affect site j. For fixed i, the $Z_{j,C}^{(i)}$, $j \in C$, are dependent Generalized Pareto random variables and hence $Z_{j,C}$, $j \in C$, are dependent Generalized Extreme Value random variables. However, we take $Z_{j,C}|\alpha_C$, $j \in C$ to be independent. From Feller (1971, Chap. 13, Sec. 6) and (3.24) this implies that the conditional distribution can be taken as

$$\Pr(Z_{j,C} < z|\alpha_C) = \exp(-\alpha_C[\tau_C\{1 + \xi_j(z - t_j)/\sigma_j\}^{-1/\xi_j}]^{q_C}). \tag{3.26}$$

Therefore, given the covariate information, the maximum of each type of extreme spatial storm has a Generalized Extreme Value distribution with parameters different from those of unconditional distribution, (3.24). From univariate extreme value theory, (3.26) is a highly realistic form for the conditional distribution.

The joint df, G, for the X's cannot immediately be obtained, but conditionally on the total unrecorded covariate information the X's are independent. From (3.25) and (3.26) we have

$$\prod_{j=1}^{p} \Pr(X_j < x_j | \alpha_C, C \in B) = \exp\left[-\sum_{C\in B} \alpha_C \tau_C^{q_C} \sum_{j\in C} \{1+\xi_j(x_j-t_j)/\sigma_j\}^{-q_C/\xi_j}\right].$$

Now, integrating over α_C for all $C \in B$ gives

$$G(x_1,\ldots,x_p) = \exp\left(-\sum_{C\in B} \tau_C \left[\sum_{j\in C}\{1+\xi_j(x_j-t_j)/\sigma_j\}^{-q_c/\xi_j}\right]^{1/q_C}\right).$$

Letting $Y_j = (\sum \tau_C)^{-1}\{1+\xi_j(X_j-t_j)/\sigma_j\}^{1/\xi_j}$, where the summation is over $C \in B_{(j)}$, the marginal distribution of Y_j is unit Fréchet for $j=1,\ldots,p$. Also, $Y_1,\ldots,Y_p$ have joint df

$$G_*(y_1,\ldots,y_p) = \exp\left[-\sum_{C\in B}\left\{\sum_{j\in C}(\psi_{j,C}/y_j)^{q_C}\right\}^{1/q_C}\right], \tag{3.27}$$

where $q_C \geq 1$ and $\psi_{j,C} = \tau_C/\sum\tau_C$, the summation being over $C \in B_{(j)}$. With $\psi_{j,C} = 0$ if $j \notin C$, then for $j = 1,\ldots,p$, $0 \leq \psi_{j,C} \leq 1$ and $\sum_{C\in B}\psi_{j,C} = 1$. It can be shown easily that $\psi_{j,C}$ is the probability that the maximum value at site j is due to a spatial storm of the type that occurs only at the collection C of sites.

The derivation of (3.27) shows that it is a valid joint df. As (3.27) satisfies (3.17) for $\alpha_{n,j} = n$ and $\beta_{n,j} = 0$, it follows that (3.27) is a multivariate extreme value distribution with unit Fréchet margins. Applying (3.9) to (3.27), we have the associated measure densities:

$$h_{m,c}(\mathbf{w}) = \left\{\prod_{k=1}^{m-1}(kq_c - 1)\right\}\left(\prod_{k\in c}\psi_{k,c}\right)^{q_c}\left(\prod_{k\in c} w_k\right)^{-(q_c+1)}\left\{\sum_{k\in c}\left(\frac{\psi_{k,c}}{w_k}\right)^{q_c}\right\}^{1/q_c - m}$$

which have $2^{p-1}(p+2)-(2p+1)$ parameters. Thus, there is mass in the interior of S_p and on each lower dimensional boundary. For $p = 2$, this distribution reduces to (3.20), the bivariate logistic distribution; hence, (3.27) is indeed a multivariate extension of the logistic distribution. Special cases of the distribution include those in Marshall and Olkin (1967), Johnson and Kotz (1972, Chap. 41) and McFadden (1978), which are obtainable as limits of (3.27) as $q_C \to \infty$ for all $C \in B$. In addition, by letting only certain $q_C \to \infty$, (3.27) can then handle cases where only some variables have singular components to their dependence structure.

Setting $\psi_{j,c_p} = 1$, $j = 1,\ldots,p$ and $q_{c_p} = q$ into (3.27), we have the symmetric logistic distribution:

$$G_*(y_1,\ldots,y_p) = \exp\{-(y_1^{-q}+\cdots+y_p^{-q})^{1/q}\}. \tag{3.28}$$

Because of its simplicity, this distribution has been studied extensively. Its characteristic function has been given by Shi (1995a). After transforming margins of G_* to Generalized Extreme Value with parameters $(\mu_j, \sigma_j, 0)$,

$$E(e^{i\mathbf{t}^T\mathbf{y}}) = e^{i\mathbf{t}^T\boldsymbol{\mu}} \frac{\Gamma(1 - i\mathbf{t}^T\boldsymbol{\sigma})}{\Gamma(1 - iq^{-1}\mathbf{t}^T\boldsymbol{\sigma})} \prod_{j=1}^{p} \Gamma(1 - iq^{-1}t_j\sigma_j),$$

where $\mathbf{y}$, $\mathbf{t}$, $\boldsymbol{\mu}$, $\boldsymbol{\sigma}$ denote p-dimensional vectors with jth components y_j, t_j, μ_j and σ_j respectively. Shi also computes the product moments of the distribution using properties of the characteristic function. Letting μ_{abcd} denote the $(a+b+c+d)$th order moment

$$\begin{aligned}
\mu_{abcd} &= E(Y_j - EY_j)^a(Y_k - EY_k)^b(Y_l - EY_l)^c(Y_m - EY_m)^d,\\
E(Y_j) &= \mu_j + \gamma\sigma_j,\\
\mu_{2000} &= \frac{\pi^2\sigma_j^2}{6},\\
\mu_{1100} &= \frac{\sigma_j\sigma_k(q^2-1)\pi^2}{6q^2},\\
\mu_{3000} &= 2\sigma_j^3\eta_3,\\
\mu_{2100} &= \frac{2\sigma_j^2\sigma_k(q^3-1)\eta_3}{q^3},\\
\mu_{1110} &= \frac{2\sigma_j\sigma_k\sigma_l(q^3-1)\eta_3}{q^3},\\
\mu_{4000} &= \frac{3\sigma_j^4\pi^4}{20},\\
\mu_{3100} &= \frac{\sigma_j^3\sigma_k(9q^2+4)(q^2-1)\pi^4}{60q^4},\\
\mu_{2200} &= \frac{\sigma_j^2\sigma_k^2(27q^4-20q^2-2q^4)\pi^4}{180q^4},\\
\mu_{2110} &= \frac{\sigma_j^2\sigma_k\sigma_l(27q^2+2)(q^2-1)\pi^4}{180q^4},\\
\mu_{1111} &= \frac{\sigma_j\sigma_k\sigma_l\sigma_m(9q^2-1)(q^2-1)\pi^4}{60q^4}
\end{aligned}$$

where $\gamma = 0.5772\ldots$ is the Euler's constant, and $\eta_s = \sum_{k=1}^{\infty} 1/k^s$ is the Zeta function (see Abramowitz and Stegun (1964)). Some special values of the Zeta function are $\eta_2 = \pi^2/6$, $\eta_3 = 1.20205690$, $\eta_4 = \pi^2/90$ and $\eta_5 = 1.03692776$. It follows that the correlation coefficient between Y_j and Y_k is $(q^2-1)/q^2$. Here the parameter q represents the amount of dependence between the two variables and has a simple interpretation as $1/q = 1-\tau$, where τ is Kendall's coefficient of concordance.

Shi (1995b) derives the Fisher information matrix of the symmetric logistic distribution (3.28), assuming Generalized Extreme Value margins with parameters (μ_j, σ_j, ξ_j). For a single observation from the distribution, the log-likelihood function is:

$$l = \log \frac{\partial^p G(x_1, \ldots, x_p)}{\partial x_1 \cdots \partial x_p}$$

$$= -\sum_{j=1}^{p} \log \sigma_j + \sum_{j=1}^{p} (q + \xi_j) \log u_j + (1 - pq) \log z - z + \log Q_p(z, q),$$

where

$$u_j = \left(1 + \xi_j \frac{x_j - \mu_j}{\sigma_j}\right)^{-1/\xi_j} \; (j = 1, \ldots, p), \; z = \left\{\sum_{j=1}^{p} u_j^q\right\}^{1/q}$$

and $Q_p(z, q)$ is a $(p-1)$th order polynomial in z satisfying:

$$Q_p(z, q) = \{q(p-1) - 1 + z\} Q_{p-1}(z, q) - z \partial Q_{p-1}(z, q)/\partial z, \; Q_1(z, q) = 1.$$

The derivation of the matrix uses the result that z is distributed independently of $\{(z^{-1}u_1)^q, \ldots, (z^{-1}u_p)^q\}$ according to a mixed gamma distribution with pdf

$$\frac{q^{1-p}}{(p-1)!} Q_p(z, q) e^{-z}, \quad z > 0.$$

Define

$$V_1(\xi; p) = E\left\{Z^{-\xi} \left(\frac{\partial}{\partial Z} \log Q_p(Z, q)\right)^2\right\},$$

$$V_2(\xi; p) = E\left\{Z^{-\xi} \frac{\partial}{\partial Z} \log Q_p(Z, q) \frac{\partial}{\partial q} \log Q_p(Z, q)\right\},$$

$$V_3(\xi; p) = E\left\{Z^{-\xi} \left(\frac{\partial}{\partial q} \log Q_p(Z, q)\right)^2\right\}$$

and introduce the notation

$$b_k = \frac{\Gamma(p + kq^{-1})}{\Gamma(1 + q^{-1})}, \quad c_k = \Psi(p + kq^{-1}) - \Psi(1 + kq^{-1}),$$

$$e_k = \Psi(p + k) - \Psi(1 + k), \quad f_k = \Psi'(p + k) - \Psi'(1 + k),$$

where $\Psi(r) = d \log \Gamma(r)/dr$ denotes the digamma function. The entire Fisher information matrix is too complicated. But the first four elements can be written as:

$$E\left(\frac{\partial l}{\partial q}\right)^2 = \frac{1}{q^4} A_0(p),$$

$$E\left(\frac{\partial l}{\partial q} \frac{\partial l}{\partial \mu_j}\right) = -\frac{1}{q^2 \sigma_j} A_1(\xi_j; p),$$

$$E\left(\frac{\partial l}{\partial q} \frac{\partial l}{\partial \sigma_j}\right) = \frac{1}{q^2 \sigma_j \xi_j} \{A_1(\xi_j; p) - A_1(0; p)\},$$

$$E\left(\frac{\partial l}{\partial q} \frac{\partial l}{\partial \xi_j}\right) = \frac{1}{q^2 \xi_j^2} \{A_1(0; p) - A_1(\xi_j; p)\} - \frac{1}{q^2 \xi_j} A_1'(0; p),$$

where

$$A_0(p) = (pq)^2\{e_0^2 - f_0 - 2(e_1^2 - f_1)\} + \left\{e_2^2 - f_2 + \frac{p-1}{p+1}\left(e_2 + \frac{3}{2} - \frac{\pi^2}{6}\right)\right\}$$
$$\times R_1(0;p) - p(p-1)q^2\left(4e_0 + \frac{2}{p} - \frac{\pi^2}{6}\right) + V_3(0;p) + 2e_1R_2(0;p),$$

$$A_1(\xi;p) = q(q+\xi)\Gamma(1+\xi)\left[(1-p)\left\{\gamma + \Psi\left(2+\frac{\xi}{q}\right)\right\} - pe_{1+q^{-1}\xi}\right]$$
$$- \mathrm{Beta}\left(p, 2+\frac{\xi}{q}\right)\left\{\left[(p-1)\left\{1-\gamma-\Psi\left(2+\frac{\xi}{q}\right)\right\}\right.\right.$$
$$\left.\left. - \left(p+1+\frac{\xi}{q}\right)e_{1+q^{-1}\xi}\right]R_1(\xi;p) - \left(p+1+\frac{\xi}{q}\right)R_2(\xi;p)\right\},$$

$$R_1(\xi;p) = \frac{\Gamma(1+\xi)P_1(\xi;p)}{\Gamma(p)} + V_1(2+\xi;p),$$

$$R_2(\xi;p) = \frac{\Gamma(1+\xi)P_2(\xi;p)}{\Gamma(p)} + V_2(1+\xi;p),$$

$$P_1(\xi;p) = (1+\xi)(2+\xi)(2b_{1+\xi} - b_{2+\xi}) - (1-pq)(1+pq+2\xi)b_\xi,$$

$$P_2(\xi;p) = (1-pq)\{(1-p)q + \xi c_\xi\} - (1+\xi)\{(1-p)q + (1+\xi)c_{1+\xi}\}b_{1+\xi}.$$

Here $A_1'(0;p)$ denotes $\partial A_1(\xi;p)/\partial\xi$ evaluated at $\xi = 0$ and so on. Obviously, these results apply only when $q > 1$ and all the $\xi_j < 1/2$.

3.5.2 *Two-Level Logistic Distributions* (Tawn, 1990)

Here we discuss a further generalization of the logistic distribution (3.27). Again we motivate it physically, following the same terminology.

It is possible that for a spatial storm that affects the collection C of sites, the values at a subset of sites D, $D \subset C$, will be more dependent than at others. Thus, sites in D may be relatively closely grouped. Then two-stage conditioning is required: the first stage represents coarse information sufficient to account for dependence between relatively widely spaced sites in $C\backslash D$, and the second stage represents finer covariate information which accounts for dependence within D. Hence, we first condition on α_C which is taken to give conditional independence within $C\backslash D$ and between $C\backslash D$ and D, but leaves D conditionally dependent. We then condition on $\alpha_{D,C}$ to give conditional independence within D.

For each $j \in C$, $C \in B$ and $D \subset C$ let $Z_{j,D,C}$ be the size of the maximum, of $N_{D,C}$ observations at site j, of storms which affect the collection C of sites where a stronger dependence exists between sites of the subset D. Here $N_{D,C}$ is taken to be Poisson with mean $\tau_{D,C}$. Hence if α is the total covariate information and α_C and $\alpha_{D,C}$ are independent positive stable variables with respective characteristic exponents $0 < 1/q_C \le 1$ and $0 < 1/q_{D,C} \le 1$, we have for $D \subset C$

$$\Pr(Z_{j,D,C} < z|\alpha) = \begin{cases} \exp\left(-\alpha_C\tau_{D,C}^{q_C}a_j^{q_C}\right) & (j \in C\backslash D), \\ \exp\left\{-\alpha_{D,C}(\alpha_C\tau_{D,C}^{q_C}a_j^{q_C})^{q_{D,C}}\right\} & (j \in D), \end{cases}$$

where $a_j = \{1 + \xi_j(z - t_j)/\sigma_j\}^{-1/\xi_j}$. In each case this distribution is Generalized Extreme Value. In the most general case, we are interested in the joint distribution of $X_1, \ldots, X_p$ where, for $j = 1, \ldots, p$,

$$X_j = \max_{C \in B_{(j)}} \max_{D \in C^*} Z_{j,D,C}$$

and D is an index variable over the set C^*, the class of all nonempty subsets of C. Then, by an analogous approach to the derivation of (3.27), the joint distribution of $X_1, \ldots, X_p$ is

$$G(x_1, \ldots, x_p) = \exp\left(- \sum_{C \in B} \left[\sum_{D \in C^*} \tau_{D,C}^{q_C} \left\{ \sum_{j \in C \setminus D} a_j^{q_C} + \left(\sum_{j \in D} a_j^{q_C q_{D,C}} \right)^{1/q_{D,C}} \right\} \right]^{1/q_C} \right).$$

Transformation of the margins to unit Fréchet form give a multivariate extreme value distribution with joint df

$$G_*(y_1, \ldots, y_p) = \exp\left\{ - \sum_{C \in B} \left(\sum_{D \in C^*} \left[\sum_{j \in C \setminus D} (\psi_{j,D,C}/y_j)^{q_C} \right.\right.\right.$$

$$\left.\left.\left. + \left\{ \sum_{j \in D} (\psi_{j,D,C}/y_j)^{q_C q_{D,C}} \right\}^{1/q_{D,C}} \right] \right)^{1/q_C} \right\}, \tag{3.29}$$

where $q_C, q_{D,C} \geq 1$ and

$$\psi_{j,D,C} = \tau_{D,C} \Big/ \left\{ \sum_{C \in B_{(j)}} \left(\sum_{D \in C^*} \tau_{D,C}^{q_C} \right)^{1/q_C} \right\}, \tag{3.30}$$

so $0 \leq \psi_{j,D,C} \leq 1$ and from (3.30) the ψ's satisfy a summation restriction. If $q_{D,C} = 1$ for all $D \subset C$, then (3.29) reduces to (3.27). Other special cases include distributions in McFadden (1978) and Joe and Hu (1996, Sec. 5.2). Because of the hierarchical form of the conditioning we call (3.29) the multivariate two-level logistic distribution. Clearly, in theory it is possible to extend this distribution to any hierarchical level.

A special case of (3.29) that has been studied to a great extent is the nested logistic distribution (Coles and Tawn, 1991) with df

$$G_*(y_1, y_2, y_3) = \exp\left[-\{(y_1^{-qq_*} + y_2^{-qq_*})^{1/q_*} + y_3^{-q}\}^{1/q}\right], \tag{3.31}$$

where $q, q_* \geq 1$. Note that the symmetric logistic distribution in (3.28) is a special case of this for $q_* = 1$. After transforming margins of (3.31) to Generalized Extreme Value with parameters $(\mu_j, \sigma_j, 0)$, Shi and Zhou (1999) derive its characteristic function as

$$E(e^{i\mathbf{t}^T\mathbf{y}}) = e^{i\mathbf{t}^T\boldsymbol{\mu}} \frac{\Gamma(1 - i\mathbf{t}^T\boldsymbol{\sigma})}{\Gamma(1 - iq^{-1}\mathbf{t}^T\boldsymbol{\sigma})} \frac{\Gamma[1 - iq^{-1}(t_1\sigma_1 + t_2\sigma_2)]}{\Gamma[1 - iq^{-1}q_*^{-1}(t_1\sigma_1 + t_2\sigma_2)]}$$

$$\times \Gamma(1 - iq^{-1}q_*^{-1}t_1\sigma_1)\Gamma(1 - iq^{-1}q_*^{-1}t_2\sigma_2)\Gamma(1 - iq^{-1}t_3\sigma_3),$$

where $\mathbf{y}$, $\mathbf{t}$, $\boldsymbol{\mu}$, $\boldsymbol{\sigma}$ denote 3-dimensional vectors with jth components y_j, t_j, μ_j and σ_j respectively. Shi and Zhou also compute the product moments of (3.31) using properties of the characteristic function. Let

$$\mu_{jkl} = E(Y_1 - EY_1)^j (Y_2 - EY_2)^k (Y_3 - EY_3)^l.$$

The second order moments are found to be:

$$\mu_{110} = \frac{\sigma_1\sigma_2(q^2q_*^2 - 1)\pi^2}{6q^2q_*^2},$$

$$\mu_{101} = \frac{\sigma_1\sigma_3(q^2 - 1)\pi^2}{6q^2},$$

$$\mu_{011} = \frac{\sigma_2\sigma_3(q^2 - 1)\pi^2}{6q^2}.$$

Thus, the correlation coefficient between Y_1 and Y_2 is $(q^2q_*^2 - 1)/(q^2q_*^2)$ and that between Y_1 and Y_3 (or Y_2 and Y_3) is $(q^2 - 1)/q^2$. The third order moments are:

$$\mu_{jkl} = 2\sigma_1^j\sigma_2^k\sigma_3^l(1 - \lambda^3)\eta_3, \quad j + k + l = 3, \quad 0 \leq j, k, l \leq 3.$$

Here $\eta_s = \sum_{k=1}^{\infty} 1/k^s$ is the Zeta function and λ takes the following values: 0 if there are two zeros among j, k, l; $1/q$ if $l \neq 0$ and one or both of j, k are nonzero; and, $1/(qq_*)$ if $l = 0$ and both of j, k are nonzero. Similarly, some fourth order moments are:

$$\mu_{310} = \mu_{130} = \frac{\sigma_1^j\sigma_2^k\sigma_3^l(9q^2q_*^2 + 4)(q^2q_*^2 - 1)\pi^4}{60q^4q_*^4}$$

and

$$\mu_{301} = \mu_{031} = \mu_{103} = \mu_{013} = \frac{\sigma_1^j\sigma_2^k\sigma_3^l(9q^2 + 4)(q^2 - 1)\pi^4}{60q^4}.$$

3.5.3 *Negative Logistic Distributions* (Joe, 1990)

This has joint df

$$G_*(y_1, \ldots, y_p) = \exp\left[-\sum_{j=1}^{p}\frac{1}{y_j} + \sum_{c\in C:|c|\geq 2}(-1)^{|c|}\left\{\sum_{j\in c}\left(\frac{\psi_{j,c}}{y_j}\right)^{q_c}\right\}^{1/q_c}\right]$$

with parameter constraints given by $q_c \leq 0$ for all $c \in C$, $\psi_{j,c} = 0$ if $j \notin c$, $\psi_{j,c} \geq 0$ for all $c \in C$ and $\sum_{c\in C}(-1)^{|c|}\psi_{j,c} \leq 1$. Again by (3.9),

$$h_{m,c}(\mathbf{w}) = \sum_{d\in C:c\subset d}(-1)^{|d|}\left\{\prod_{k=1}^{m-1}(1 - kq_d)\right\}\left(\prod_{k\in c}\psi_{k,d}\right)^{q_d}\left(\prod_{k\in c}w_k\right)^{-(q_d+1)}$$
$$\times\left\{\sum_{k\in c}\left(\frac{\psi_{k,d}}{w_k}\right)^{q_d}\right\}^{1/q_d - m}.$$

Evidently this has structure similar to the logistic family with the special case for $\psi_{j,c_p} = 1$, $j = 1, \ldots, p$ and $q_{c_p} = q$ giving a symmetric version that has all its mass in the interior of S_p.

The two bilogistic families discussed earlier are asymmetric generalizations of the logistic families, but it is not yet known how they generalize to the multivariate case. However, the family of Beta distributions generalizes to the following.

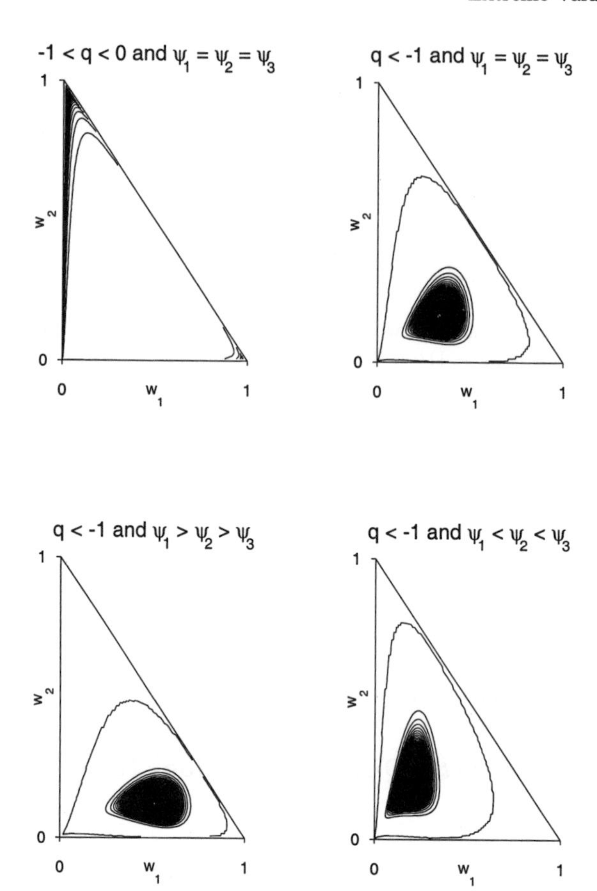

Figure 3.8: Possible forms for $h(w_1, w_2)$ for the trivariate negative logistic distribution.

3.5.4 *Dirichlet Distributions* (Coles and Tawn, 1991)

The pdf of Dirichlet($q_1, \ldots, q_p$) distribution is

$$h_{\dagger}(\mathbf{w}) = \left\{\prod_{j=1}^{p} \Gamma(q_j)\right\}^{-1} \Gamma(q_1 + \cdots + q_p) \prod_{j=1}^{p} w_j^{q_j - 1}, \; q_j > 0, j = 1, \ldots, p, \mathbf{w} \in S_p.$$

By eq. (3.10) $m_j = q_j/(q_1 + \cdots + q_p)$, and from eq. (3.11) it follows that

$$h_{p,c_p}(\mathbf{w}) = \frac{\Gamma(q_1 + \cdots + q_p + 1)}{(q_1 w_1^* + \cdots + q_p w_p^*)^{p+1}} \prod_{j=1}^{p} \frac{q_j}{\Gamma(q_j)} \prod_{j=1}^{p} \left(\frac{q_j w_j}{q_1 w_1^* + \cdots + q_p w_p^*}\right)^{q_j - 1}$$

is the density, in the interior of S_p, of a valid measure H_* that satisfies the constraints (3.7). This has structure similar to the Beta family, the special case for $p = 2$, with symmetry arising when $q_1 = \cdots = q_p$.

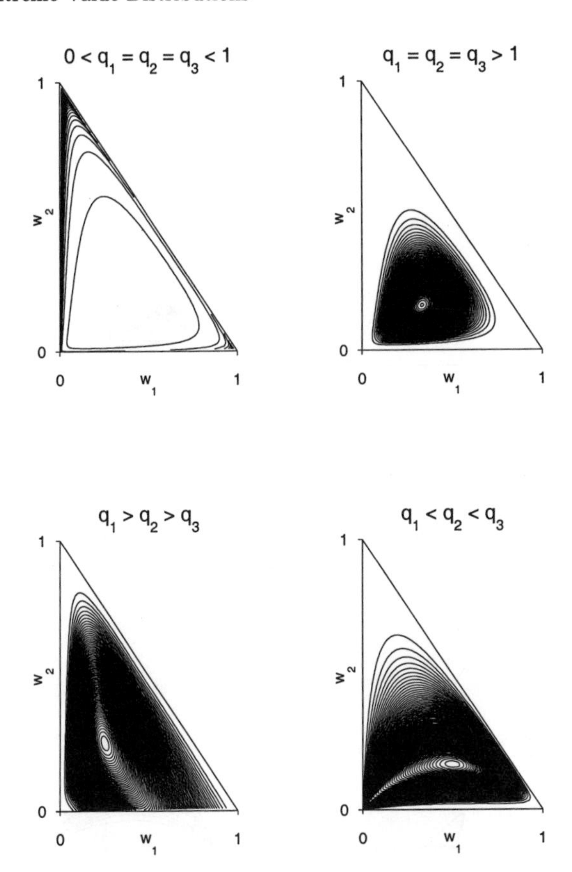

Figure 3.9: Possible forms for $h(w_1, w_2)$ for the Dirichlet distribution.

The corresponding form for G_* is complicated although numerical computation is feasible. This distribution has been found most suitable for estimating in continuous space the spatial dependence within rainfall storms (Coles, 1993) and for estimating the dependence between the extremes of surge, wave height and wave period (Coles and Tawn, 1994).

3.5.5 *Time Series Logistic Distributions* (Coles and Tawn, 1991)

Let $Y_1, \ldots, Y_p$ be a first-order Markov process representing a time series such as observations of a propagating sea storm at sites ordered along a coast. Suppose without loss of generality that Y_j have the unit Fréchet distribution. Let $f^{(j)}$ denote the joint density of (Y_j, Y_{j+1}). Then the joint density of $Y_1, \ldots, Y_p$ is

$$f(y_1, \ldots, y_p) = \Phi_1'(y_1) \prod_{j=1}^{p-1} \frac{f^{(j)}(y_j, y_{j+1})}{\Phi_1'(y_j)}.$$

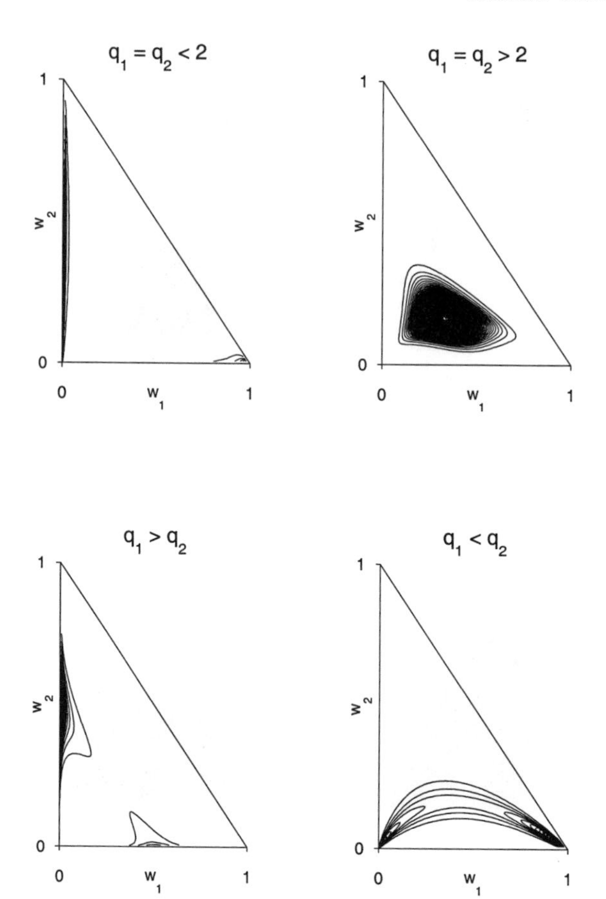

Figure 3.10: Possible forms for $h(w_1, w_2)$ for the time series logistic distribution.

Evaluating (3.16) at $B = [0, y_1] \times \cdots \times [0, y_p]$ and then differentiating it with respect to $y_1, \ldots, y_p$, we have

$$\lim_{t \to \infty} \{t^{p+1} f(ty_1, \ldots, ty_p)\} = \left(\sum_{j=1}^{p} y_j \right)^{-(p+1)} h \left\{ \frac{y_1}{\sum_{j=1}^{p} y_j}, \ldots, \frac{y_p}{\sum_{j=1}^{p} y_j} \right\}.$$

Hence if we assume that the joint df of (Y_j, Y_{j+1}) belongs to the domain of attraction of a logistic bivariate extreme value df with $\psi_1 = \psi_2 = 1$ and $q_j = q$ then

$$h(\mathbf{w}) = \frac{1}{w_1^2} \prod_{j=1}^{p-1} \frac{(q_j - 1) w_j^2}{(w_j w_{j+1})^{q_j + 1}} (w_j^{-q_j} + w_{j+1}^{-q_j})^{1/q_j - 2}, \quad \mathbf{w} \in S_p$$

for $q_j \geq 1$.

An extension of this family, not examined here, is based on a higher order Markov sequence with the associated joint density of consecutive values taken as multivariate extreme value with unit Fréchet margins.

3.5.6 *Distributions Closed Under Margins*

Here we discuss a few technical tools for constructing multivariate extreme value distributions that are closed under margins. We begin with one communicated by Nadarajah (1999c).

Let G_A, G_B and G_C be known bivariate extreme value distributions with respective exponent measure functions V_A, V_B and V_C. Assume as usual that all univariate margins are unit Fréchet. The following steps construct a trivariate extreme value distribution that has G_A, G_B and G_C as its bivariate margins.

Partition the two-dimensional unit simplex, S_3, into three disjoint sets B_j, $j = 1, 2, 3$ chosen as

$$\begin{aligned} B_1 &= \{(w_1, w_2) \in S_3 : w_1 \geq w_2 \text{ and } 2w_1 + w_2 \geq 1\}, \\ B_2 &= \{(w_1, w_2) \in S_3 : w_2 \geq w_1 \text{ and } 2w_2 + w_1 \geq 1\}, \\ B_3 &= \{(w_1, w_2) \in S_3 : 2w_1 + w_2 \leq 1 \text{ and } 2w_2 + w_1 \leq 1\}. \end{aligned}$$

Define $H_j : B_j \to \Re_+$ and $\theta_j : \Re_+^3 \to S_3$ to satisfy

$$\begin{aligned} \int_{B_1} w_1 dH_1(w_1, w_2) &= \int_{B_2} w_2 dH_2(w_1, w_2) \\ &= \int_{B_3} (1 - w_1 - w_2) dH_3(w_1, w_2) = 1, \end{aligned}$$

$$\begin{aligned} &\sum_{j=1}^{3} \theta_j(y_1, y_2, y_3) = 1, \\ &\theta_j(ky_1, ky_2, ky_3) = \theta_j(y_1, y_2, y_3), \quad j = 1, 2, 3, \qquad k > 0, \\ &\quad \theta_1(y_1, y_2, \infty) = \theta_2(y_1, \infty, y_3) = \theta_3(\infty, y_2, y_3) = 1. \end{aligned}$$

Then,

$$G_*(y_1, y_2, y_3) = \exp\left\{ -\sum_{j=1}^{3} \theta_j(y_1, y_2, y_3) V_j(y_1, y_2, y_3) \right\},$$

where

$$V_j(y_1, y_2, y_3) = \int_{B_j} \max[w_1 V_A(y_1, y_2), w_2 V_B(y_1, y_3), (1 - w_1 - w_2) V_C(y_2, y_3)] dH_j(w_1, w_2),$$

is a trivariate extreme value distribution. It is easily checked that

$$G_*(y_1, y_2, \infty) = G_A(y_1, y_2), \quad G_*(y_1, \infty, y_3) = G_B(y_1, y_3) \text{ and } G_*(\infty, y_2, y_3) = G_C(y_2, y_3).$$

The second construct that we discuss is due to Marco and Ruiz–Rivas (1992). Suppose that we can express

$$G_j = \nu_j(F_j), \qquad j = 1, 2$$

where F_j is an n_j-variate df and $\nu_j : [0,1] \to [0,1]$ is continuous to the right and satisfies:

$$\nu_j(0) = 0, \quad \nu_j(1) = 1, \quad \nu_j^{('k)} \geq 0, \quad 1 \leq k \leq n_j.$$

Let

$$a_{jk} = \nu_j^{('k)}(0)/k!,\ k = 1, \ldots, n_j - 1, j = 1, 2,\ c_j = 1 - \sum_{k=1}^{n_j - 1} a_{jk}.$$

Define

$$\nu(s_1, s_2) = \int_0^{s_1} \int_0^{s_2} \left(\frac{s_1 - u_1}{1 - u_1}\right)^{n_1 - 1} \left(\frac{s_2 - u_2}{1 - u_2}\right)^{n_2 - 1} dD(u_1, u_2),$$

where $D(u_1, u_2)$ is any df in $[0,1]^2$ with marginal densities:

$$\rho_j(u_j) = \frac{1}{c_j} \frac{(1 - u_j)^{n_j - 1}}{(n_j - 1)!} \nu_j^{('n_j)}(u_j), \quad u_j \in [0,1].$$

It is then easily verified that

$$G(\mathbf{x}_1, \mathbf{x}_2) = \nu(F_1(\mathbf{x}_1), F_2(\mathbf{x}_2))$$

is a df with G_1, G_2 as marginals.

We now discuss three more specialized constructs (due to Joe (1994)) that are closed under margins. Let V_m denote the exponent measure function of an m-variate extreme value distribution. Take $V_2(y_1, y_2) = (y_1^{-q} + y_2^{-q})^{1/q}$ and, for $m \geq 3$, define V_m to satisfy the recurrence relation

$$V_m(y_1, \ldots, y_m) = [\{V_{m-1}(y_1, \ldots, y_{m-1})\}^{q_{1,m}} + y_m^{-q_{1,m}}]^{1/q_{1,m}}, \tag{3.32}$$

where $q_{1,2} \geq q_{1,3} \geq \cdots q_{1,m} \geq 1$. Then V_m corresponds to a multivariate logistic distribution that is closed under margins. With V_k as given by (3.32), define

$$V_m^*(y_1, \ldots, y_m) = \frac{1}{y_1} + \cdots + \frac{1}{y_m} - \sum_{j_1 < j_2} (y_{j_1}^{-q_{1,j_2}} + y_{j_2}^{-q_{1,j_2}})^{1/q_{1,j_2}}$$
$$+ \sum_{3 \leq k \leq m} (-1)^{k+1} \sum_{j_1 < \cdots < j_k} V_k(y_{j_1}, \ldots, y_{j_k})$$

where $q_{1,2} \leq \cdots \leq q_{1,m} \leq 0$. Then V_m^* corresponds to a multivariate negative logistic distribution that is closed under margins. For the final construct, take

$$V_2(y_1, y_2) = \frac{1}{y_1} \Phi\left(\frac{a}{2} + \frac{1}{a} \log \frac{y_2}{y_1}\right) + \frac{1}{y_2} \Phi\left(\frac{a}{2} + \frac{1}{a} \log \frac{y_1}{y_2}\right),$$

the exponent measure function corresponding to the bivariate extreme value Gaussian distribution (where Φ denotes the cdf of the standard Normal distribution). Let $\rho_{jkl} = (a_{kj}^2 + a_{lj}^2 - a_{kl}^2)/(2a_{kj}a_{lj})$ for j, k, l distinct and, for $m \geq 3$, define V_m recursively by

$$V_m(y_1, \ldots, y_m) = V_{m-1}(y_1, \ldots, y_{m-1})$$
$$+ \int_0^{1/y_m} \Phi\left(\frac{a_{1,m}}{2} + \frac{\log(y_1 s)}{a_{1,m}}, \ldots, \frac{a_{m-1,m}}{2} + \frac{\log(y_{m-1} s)}{a_{m-1,m}}\right) ds.$$

Here $\Phi()$ denotes the cdf of a multivariate Normal distribution with means 0, correlations $\rho_{mkl}(1 \leq k < l < m - 1)$, and variances 1. This construct ensures that V_m corresponds to a multivariate extreme value Gaussian distribution that is closed under margins.

3.6 Statistical Estimation

3.6.1 *Parametric Estimation*

As with the univariate approach based on the Generalized Extreme Value distribution any approach for modelling multivariate extremes directly based on the componentwise maxima $\mathbf{M}_n$ is wasteful of data (Coles, 1991). A further weakness is that $\mathbf{M}_n$ may not correspond to an observed event, so any approach based on $\mathbf{M}_n$ may give misleading results in small sample applications (Coles, 1991). In this section we describe two likelihood-based parametric approaches for estimation which circumvent these drawbacks.

The first approach is based on the limiting point process result (3.5) and is due to the independent efforts of Coles and Tawn (1991) and Joe *et al.* (1992). In (3.5), they assume, for large n, that $\mathcal{P}_n$ in a region B, bounded away from $\{(0,\ldots,0)\}$ by a distance dependent on the rate of convergence, is approximately a nonhomogeneous Poisson process with intensity satisfying eq. (3.6). Take $\{n^{-1}(U_1(X_{i,1}),\ldots,U_p(X_{i,p})), i=1,\ldots,n_B\}$ to be the points of $\mathcal{P}_n$ in B. Then the likelihood over B, L_B, is

$$L_B(\boldsymbol{\theta}; n^{-1}\{U_1(X_{i,1}),\ldots,U_p(X_{i,p})\}) = \exp\{-\mu(B)\} \prod_{i=1}^{n_B} r_i^{-2} dr H_*(d\mathbf{w}_i), \tag{3.33}$$

where $\boldsymbol{\theta}$ are parameters for H_* and $(r_i, \mathbf{w}_i) = T(U_1(X_{i,1}),\ldots,U_p(X_{i,p}))$ are the pseudo-polar transforms given by (3.4).

Now consider the general likelihood for $\{\mathbf{X}_i, i=1,\ldots,n\}$. This involves simultaneous estimation of the marginal parameters for U_j and the dependence parameters for H_*. Hence we require an appropriate choice for the region B and a model for the marginal transformations U_j to be included in eq. (3.33).

Coles and Tawn (1991) and Joe *et al.* (1992) regard the region $B = \Re_+^p \backslash \{(0,u_1)\times\cdots\times(0,u_p)\}$ for high thresholds u_j as a sensible choice as it contains all observations which are large in at least one margin. It also ensures that points in B are invariant to the model for marginal transformations, (3.34), chosen below.

The marginal transformations above a high threshold, t_j (say), are determined by the conditional distribution of threshold exceedances. As noted in Chap. 2 these have a Generalized Pareto distribution form: $\Pr(X_j > x | X_j > t_j) = \{1+\xi_j(x-t_j)/\sigma_j\}^{-1/\xi_j}$, $\sigma_j > 0$, $1+\xi_j(x-t_j)/\sigma_j > 0$. Thus, for $X_j > t_j$

$$\Pr(X_j > x) = \lambda_j\{1+\xi_j(x-t_j)/\sigma_j\}^{-1/\xi_j}.$$

Here $\lambda_j = \Pr(X_j > t_j)$ is obtained as the proportion of points exceeding t_j. Points below the threshold are relatively dense, so we transform these components using the empirical df. Hence, the model for the marginal transformation is:

$$U_j(X_j) = \begin{cases} -(\log[1-\lambda_j\{1+\xi_j(X_j-t_j)/\sigma_j\}^{-1/\xi_j}])^{-1} & \text{if } X_j > t_j \\ -[\log\{R(X_j)/(n+1)\}]^{-1} & \text{if } X_j \le t_j \end{cases} \tag{3.34}$$

where $R(X_j)$ denotes the rank of X_j.

Hence, the thresholds for the limiting process are given by $u_j = n^{-1}U_j(X_j)$, though checks are required to ensure that these are sufficiently high for eq. (3.6) to be valid in B — see Nadarajah (1994, Chap. 4) for diagnostics that ensure this. Incorporating eqs. (3.6) and

(3.34) into likelihood function (3.33) gives, for Cartesian components, the likelihood function $L_B(\boldsymbol{\theta}, \boldsymbol{\sigma}, \boldsymbol{\xi}; \{\mathbf{X}_i\})$ as

$$\exp\{-V(u_1,\ldots,u_p)\}\prod_{i=1}^{n_B} h(\mathbf{w}_i)(nr_i)^{-(p+1)}$$

$$\times \prod_{i=1}^{n_B} \prod_{\substack{j=1,\ldots,p \\ U_j(X_{i,j})>nu_j}} [\sigma_j^{-1}\lambda_j^{-\xi_j} U_j^2(X_{i,j}) \exp(1/U_j(X_{i,j}))\{1-\exp(-1/U_j(X_{i,j}))\}^{1+\xi_j}],$$

where $\boldsymbol{\sigma} = (\sigma_1,\ldots,\sigma_p)$, $\boldsymbol{\xi} = (\xi_1,\ldots,\xi_p)$ and $h(\mathbf{w})$ is a parametric form for the density of H_* in the interior of S_p.

Generally maximum likelihood estimators behave regularly provided that marginally $\xi_j > -1/2$, $j = 1,\ldots,p$ (Smith, 1985). However, in some cases dependence parameters are super-efficient. For example, for the symmetric logistic distribution there is discontinuity in h as $q \downarrow 1$: thus within the logistic distributions any vertex mass implies $q = 1$, corresponding to independence of the variables. Tawn (1988b, 1990) discusses this problem further.

Ledford and Tawn (1996) and Smith *et al.* (1997) have developed an alternative approach based on (3.15):

$$\lim_{t\to\infty}\frac{-\log F_*(ty_1,\ldots,ty_p)}{-\log F_*(t,\ldots,t)} = \frac{-\log G_*(y_1,\ldots,y_p)}{-\log G_*(1,\ldots,1)} = \frac{V(y_1,\ldots,y_p)}{V(1,\ldots,1)}. \tag{3.35}$$

In using this result, they assume that (3.35) holds as an identity for some fixed large $t = t_c$ say. Since $t_c y_j$ needs to be large for each j it is imposed that $y_j' = t_c y_j$ is above some high threshold. This gives

$$\log F_*(y_1',\ldots,y_p') = \log F_*(t_c,\ldots,t_c)\frac{V(y_1'/t_c,\ldots,y_p'/t_c)}{V(1,\ldots,1)}.$$

Now, by (3.8), V is a homogeneous function of order -1, and the y' terms are just dummy variables, so

$$\begin{aligned} F_*(y_1,\ldots,y_p) &= \exp\left\{V(y_1,\ldots,y_p)\frac{t_c\log F_*(t_c,\ldots,t_c)}{V(1,\ldots,1)}\right\} \\ &= \exp\{V(y_1,\ldots,y_p)K\} \end{aligned} \tag{3.36}$$

for some K, when each y_j is above some suitably high threshold. Ledford and Tawn (1996) take this threshold as the $1-\lambda_j$ quantile of the unit Fréchet distribution, where λ_j is some small probability. Thus, $y_j \geq -1/\log(1-\lambda_j)$ for $j = 1,\ldots,p$.

To evaluate the constant K, set $y_1 = -1/\log(1-\lambda_1)$, $y_2 = \cdots = y_p = \infty$ into (3.36) and note that

$$F_*\{-1/\log(1-\lambda_1),\infty,\ldots,\infty\} = \exp[V\{-1/\log(1-\lambda_1),\infty,\ldots,\infty\}K].$$

This implies that $1-\lambda_1 = \exp\{-K\log(1-\lambda_1)\}$, and so $K = -1$. Thus, combining (3.36) with the definition of F_*, we have

$$F(x_1,\ldots,x_p) = \exp[-V\{-1/\log F_1(x_1),\ldots,-1/\log F_p(x_p)\}],$$

valid for each $x_j \geq t_j$ say, where the t_j are chosen so that, for each j, $F_j(t_j) = 1-\lambda_j$. Further, assuming that F_j for $X_j \geq t_j$ has the Generalized Pareto distribution with parameters $(\sigma_j, \xi_j, \lambda_j)$, we have

$$F(x_1, \ldots, x_p) = \exp\{-V(U_1(x_1), \ldots, U_p(x_p))\}, \tag{3.37}$$

valid for each $x_j \geq t_j$, $j = 1, \ldots, p$. This derivation simply amounts to assuming that for joint exceedances of a set of suitably high thresholds, the dependence structure of the df F is that of an exact multivariate extreme value distribution. The approach due to Smith *et al.* (1997) slightly differs from this in that they assume

$$F(x_1, \ldots, x_p) = 1 - V(1/[1 - \exp\{-1/U_1(x_1)\}], \ldots, 1/[1 - \exp\{-1/U_p(x_p)\}]) \tag{3.38}$$

which is a first-order approximation of (3.37) for small λ_j.

To develop a likelihood based on (3.37), Ledford and Tawn (1996) consider marginal observations below their respective thresholds as censored at the threshold. Thus, the contribution to the likelihood of a typical point $(x_1, \ldots, x_p)$ for which the margins $j_1, \ldots, j_m$ attain or exceed their thresholds is given by

$$\frac{\partial^m F(x_1, \ldots, x_p)}{\partial x_{j_1} \cdots \partial x_{j_m}}, \tag{3.39}$$

with F given by (3.37), evaluated at $x_1 = \max(t_1, X_1), \ldots, x_p = \max(t_p, X_p)$.

We construct the explicit form of the likelihood for the case $p = 2$ as follows. For high marginal thresholds t_1 and t_2, divide the (X_1, X_2) plane into four regions based on whether each margin is above or below its respective threshold. Label these regions $B_{i_1 i_2}$, with $i_j = 0$ if $X_j < t_j$ and $i_j = 1$ if $X_j \geq t_j$, for $j = 1, 2$. Let $L_{i_1 i_2}(x_1, x_2)$ denote the likelihood expression corresponding to a point (x_1, x_2) falling in region $B_{i_1 i_2}$. Write $r_j = -1/\log(1-\lambda_j)$ and define

$$\begin{aligned}
V_1(y_1, y_2) &= -\frac{\partial}{\partial y_1} V(y_1, y_2), \\
V_2(y_1, y_2) &= -\frac{\partial}{\partial y_2} V(y_1, y_2), \\
V_{12}(y_1, y_2) &= -\frac{\partial^2}{\partial y_1 \partial y_2} V(y_1, y_2).
\end{aligned}$$

Then (3.39) gives that

$$\begin{aligned}
L_{00}(x_1, x_2) &= \exp\{-V(r_1, r_2)\}, \\
L_{01}(x_1, x_2) &= \exp[-V\{r_1, U_2(x_2)\}] V_2\{r_1, U_2(x_2)\} K_2, \\
L_{10}(x_1, x_2) &= \exp[-V\{U_1(x_1), r_2\}] V_1\{U_1(x_1), r_2\} K_1, \\
L_{11}(x_1, x_2) &= \exp[-V\{U_1(x_1), U_2(x_2)\}][V_1\{U_1(x_1), U_2(x_2)\} V_2\{U_1(x_1), U_2(x_2)\} \\
&\quad + V_{12}\{U_1(x_1), U_2(x_2)\}] K_1 K_2,
\end{aligned} \tag{3.40}$$

where

$$K_j = \lambda_j \sigma_j^{-1} \{1 + \xi_j (x_j - t_j)/\sigma_j\}_+^{-(1+\xi_j^{-1})} \{U_j(x_j)\}^2 \exp\{1/U_j(x_j)\}$$

for $j = 1, 2$. Thus, the likelihood contribution from a typical point $(x_{i,1}, x_{i,2})$ for dependence parameters $\boldsymbol{\theta}$ and marginal parameters $\boldsymbol{\sigma} = (\sigma_1, \sigma_2)$, $\boldsymbol{\xi} = (\xi_1, \xi_2)$, $\boldsymbol{\lambda} = (\lambda_1, \lambda_2)$ is given by

$$L_i(\boldsymbol{\theta}, \boldsymbol{\sigma}, \boldsymbol{\xi}, \boldsymbol{\lambda}) = \sum_{r,s \in (0,1)} L_{rs}(x_{i,1}, x_{i,2}) I_{rs}(x_{i,1}, x_{i,2}),$$

where $I_{rs}(x_{i,1}, x_{i,2}) = I\{(x_{i,1}, x_{i,2}) \in B_{rs}\}$ with I the indicator function. Writing the likelihood for a set of n independent points as $L_{(n)}()$, we thus have

$$L_{(n)}(\boldsymbol{\theta}, \boldsymbol{\sigma}, \boldsymbol{\xi}, \boldsymbol{\lambda}) = \prod_{i=1}^{n} L_i(\boldsymbol{\theta}, \boldsymbol{\sigma}, \boldsymbol{\xi}, \boldsymbol{\lambda}).$$

We noted earlier that problems arise with the point process approach when the marginal variables are independent. Nadarajah (1994, Chap. 3) shows that these problems remain for the approach due to Smith *et al.* (1997). But the approach of Ledford and Tawn (1996) overcomes these problems. To test for exact independence of the marginal extremes, they consider, for a typical point $(x_{i,1}, x_{i,2})$, the behavior of the score at independence defined by

$$\mathcal{U}_i = \frac{d}{d\boldsymbol{\theta}} \log L_i(\boldsymbol{\theta}, \boldsymbol{\sigma}, \boldsymbol{\xi}, \boldsymbol{\lambda})|_{\boldsymbol{\theta}=\boldsymbol{\theta}_I},$$

where $\boldsymbol{\theta}_I$ is the vector of dependence parameters corresponding to total independence and $(\boldsymbol{\sigma}, \boldsymbol{\xi}, \boldsymbol{\lambda})$ is the vector of the marginal parameters which jointly maximizes $L_{(n)}(\boldsymbol{\theta}_I, \boldsymbol{\sigma}, \boldsymbol{\xi}, \boldsymbol{\lambda})$. Evaluating this for the possible forms in (3.40) shows that the score depends on the marginal parameters only through the unit Fréchet variables, $U_j(x_j)$, and the transformed thresholds, r_j, for $j = 1, 2$. Thus,

$$\mathcal{U}_i = \sum_{r,s \in (0,1)} T_{rs}(x_{i,1}, x_{i,2}) I_{rs}(x_{i,1}, x_{i,2}),$$

where

$$\begin{aligned}
T_{00}(x_{i,1}, x_{i,2}) &= r_1^{-1} \log r_1^{-1} + r_2^{-1} \log r_2^{-1} - (r_1^{-1} + r_2^{-1}) \log(r_1^{-1} + r_2^{-1}), \\
T_{01}(x_{i,1}, x_{i,2}) &= r_1^{-1} \log r_1^{-1} + (u_2^{-1} - 1) \log u_2^{-1} - (r_1^{-1} + u_2^{-1} - 1) \log(r_1^{-1} + u_2^{-1}), \\
T_{10}(x_{i,1}, x_{i,2}) &= (u_1^{-1} - 1) \log u_1^{-1} + r_2^{-1} \log r_2^{-1} - (u_1^{-1} + r_2^{-1} - 1) \log(u_1^{-1} + r_2^{-1}), \\
T_{11}(x_{i,1}, x_{i,2}) &= (u_1^{-1} - 1) \log u_1^{-1} + (u_2^{-1} - 1) \log u_2^{-1} \\
&\quad - (u_1^{-1} + u_2^{-1} - 2) \log(u_1^{-1} + u_2^{-1}) - (u_1^{-1} + u_2^{-1})^{-1},
\end{aligned}$$

where $u_j = U_j(x_j)$ for $j = 1, 2$. Defining the total score of a set of n independent points as

$$\mathcal{U}_{(n)} = \sum_{i=1}^{n} \mathcal{U}_i,$$

Ledford and Tawn show, under the assumption of independence of the margins, that

$$-\mathcal{U}_{(n)}/(2^{-1} n \log n)^{1/2} \to N(0, 1)$$

as $n \to \infty$, with departures from independence producing large positive variates. They also give a table of critical points relevant for testing, obtained by simulation.

3.6.2 *Semiparametric Estimation*

The parametric approaches discussed above are based on the use of asymptotically justified approximations for both the marginal distributions, F_j, and the dependence structure in the joint tail of F. Models derived from these approximations are fitted to a region of the observed joint tail which is determined by suitably chosen high thresholds. A drawback with this is the necessity for the same thresholds to be taken for the convergence of both marginal and dependence aspects which can result in inefficient estimation. Dixon and Tawn (1995a) provide an extension which removes this constraint. The resulting model is semiparametric and requires computationally intensive techniques for likelihood evaluation.

We provide an outline of the arguments leading to the likelihood for $p = 2$. Consider the parametric model, (3.38), due to Smith *et al.* (1997). In (3.38), both the limiting marginal aspects, where

$$F_j(x_j) = 1 - \lambda_j[1 + \xi_j(x_j - t_j)/\sigma_j]_+^{-1/\xi_j}, \quad j = 1, 2, \tag{3.41}$$

and the limiting dependence aspects, where

$$F(x_1, x_2) = 1 - V(1/[1 - \exp\{-1/U_1(x_1)\}], 1/[1 - \exp\{-1/U_2(x_2)\}]), \tag{3.42}$$

are assumed to hold for all $x_j \geq t_j$. In contrast, Dixon and Tawn (1995a) assume that the dependence structure convergence is slower than the marginal convergence. They introduce thresholds for the dependence structure, t_{d_j}, such that (3.42) holds for $x_j \geq t_j = t_{d_j}$, and thresholds for the margins, $t_{m_j} (< t_{d_j})$, such that (3.41) holds for $x_j \geq t_j = t_{m_j}$. Correspondingly the sample space is partitioned into nine regions

$$B_{kl} = \{\mathbf{x} \in \Re^2 : b_{1,k-1} \leq x_1 < b_{1,k}, b_{2,l-1} \leq x_2 < b_{2,l}\}, \qquad k, l = 0, 1, 2,$$

where the boundaries are $b_{k,-1} = -\infty$, $b_{k,0} = t_{m_k}$, $b_{k,1} = t_{d_k}$ and $b_{k,2} = \infty$, $k = 1, 2$. Based on this partition, let

$$\begin{aligned} \Omega_1 &= \bigcap_{k=0}^{2} B_{2k}^c \bigcap \bigcap_{k=0}^{2} B_{k2}^c, \\ \Omega_2 &= B_{22} \cup B_{12} \cup B_{21}, \\ \Omega_3 &= B_{20} \cup B_{02}. \end{aligned}$$

Then,

- in Ω_2, (3.41) and (3.42) hold with t_j replaced by t_{m_j}. Thus, the likelihood contribution from an observation $\mathbf{x} \in \Omega_2$ is

$$L_{\Omega_2}(\mathbf{x}) = \frac{\partial^2}{\partial x_1 \partial x_2} F(\mathbf{x}) \tag{3.43}$$

with F given by (3.42).
- in Ω_3, the assumptions do not provide a complete model as only one marginal form is specified. Consequently, following Smith *et al.* (1997), the likelihood is obtained by censoring observations below the marginal thresholds. Thus, for $\mathbf{x} \in B_{20}$,

$$L_{B_{20}}(\mathbf{x}) = \frac{\partial}{\partial x_1} F(x_1, t_{m_2}) \tag{3.44}$$

and, for $\mathbf{x} \in B_{02}$,

$$L_{B_{02}}(\mathbf{x}) = \frac{\partial}{\partial x_2} F(t_{m_1}, x_2), \tag{3.45}$$

where F is given by (3.42).

- in Ω_1, as above, the joint distribution is not completely specified. A solution is developed using a combination of both parametric and censoring approaches which leads to a semiparametric model. Let $L_{\Omega_1}^{(c)}$ denote the likelihood of an observation conditional on it being in Ω_1. Then the unconditional likelihood can be written as

$$L_{\Omega_1}(\mathbf{x}) = F(t_{d_1}, t_{d_2}) L_{\Omega_1}^{(c)}(\mathbf{x}), \tag{3.46}$$

where F is given by (3.42). Let $F^{(c)}$ denote the corresponding conditional joint distribution in Ω_1. Since the marginal distributions of F are given by (3.41) above $t_j = t_{m_j}$, the marginal distributions of $F^{(c)}$ are

$$F_1^{(c)}(x_1) = F(x_1, t_{d_2}) / F(t_{d_1}, t_{d_2}), \qquad t_{m_1} \leq x_1 \leq t_{d_1} \tag{3.47}$$

with $F_2^{(c)}$ given similarly. Despite knowing its marginal distributions, we do not have an explicit model for $F^{(c)}$. Thus, it is estimated empirically with $\tilde{f}^{(c)}(\mathbf{x})$ denoting a kernel density estimate of $f^{(c)}(\mathbf{x})$, the joint density of $F^{(c)}$, and $\tilde{F}^{(c)}$ denoting the corresponding estimate of $F^{(c)}$ obtained by integrating $\tilde{f}^{(c)}$. The derivation of $L_{\Omega_1}^{(c)}$, is equivalent to the problem of obtaining the joint distribution with the required marginal form given by (3.47) and a dependence structure consistent with interaction form of a non-parametric estimate within the region. A general solution to this type of model generation problem is given by iterative proportional fitting algorithm (Bishop *et al.*, 1975; Whittaker, 1990). Given two marginal densities e_1 and e_2 and a joint density $g^{(0)}(\mathbf{x})$, the sequence of densities $g^{(k)}(\mathbf{x})$, with jth marginal $g_j^{(k)}(x_j)$, given by the two stage updating procedure

$$g^{(2k-1)}(\mathbf{x}) = g^{(2k-2)}(\mathbf{x}) e_1(x_1) / g_1^{(2k-2)}(x_1),$$
$$g^{(2k)}(\mathbf{x}) = g^{(2k-1)}(\mathbf{x}) e_2(x_2) / g_2^{(2k-1)}(x_2),$$

$k = 1, 2, \ldots$ converges pointwise to a limiting density $g^{\infty}(\mathbf{x})$. Here, $g^{\infty}(\mathbf{x})$ has marginal densities $e_j (j = 1, 2)$, and a dependence structure which is equivalent to that of $g^{(0)}(\mathbf{x})$ in the sense that

$$\partial^2 \log g^{(0)}(\mathbf{x}) / \partial x_1 \partial x_2 = \partial^2 \log g^{(\infty)}(\mathbf{x}) / \partial x_1 \partial x_2.$$

Thus, taking

$$e_j(x_j) = \begin{cases} F^{(c)}(t_{m_j}) & \text{if } x_j \leq t_{m_j} \\ \partial F_j^{(c)}(x_j) / \partial x_j & \text{if } t_{m_j} < x_j < t_{d_j}, \end{cases}$$

with $F_j^{(c)}$ given by (3.47), and

$$g^{(0)}(\mathbf{x}) = \begin{cases} \tilde{F}^{(c)}(t_{m_1}, t_{m_2}) & \text{if } \mathbf{x} \in B_{00} \\ \partial \tilde{F}^{(c)}(x_1, t_{m_2}) / \partial x_1 & \text{if } \mathbf{x} \in B_{10} \\ \partial \tilde{F}^{(c)}(t_{m_1}, x_2) / \partial x_2 & \text{if } \mathbf{x} \in B_{01} \\ \tilde{f}^{(c)}(\mathbf{x}) & \text{if } \mathbf{x} \in B_{11}, \end{cases}$$

the iterative procedure gives the conditional likelihood contribution for an observation for $\mathbf{x} \in \Omega_1$ as

$$L_{\Omega_1}^{(c)}(\mathbf{x}) = g^{(\infty)}(\mathbf{x})$$

and hence from (3.46)

$$L_{\Omega_1}(\mathbf{x}) = F(t_{d_1}, t_{d_2}) g^{(\infty)}(\mathbf{x}). \tag{3.48}$$

By combining the calculations, the likelihood for an iid sample $\{\mathbf{x}_i, i = 1, \ldots, n\}$ is given by

$$L(\boldsymbol{\theta}, \boldsymbol{\sigma}, \boldsymbol{\xi}, \boldsymbol{\lambda}) = \prod_{i=1}^{n} \{L_{\Omega_2}(\mathbf{x}_i)^{I\{\mathbf{x}_i \in \Omega_2\}} L_{B_{02}}(\mathbf{x}_i)^{I\{\mathbf{x}_i \in B_{02}\}} L_{B_{20}}(\mathbf{x}_i)^{I\{\mathbf{x}_i \in B_{20}\}} L_{\Omega_1}(\mathbf{x}_i)^{I\{\mathbf{x}_i \in \Omega_1\}}\},$$

where I is the indicator function, and L_{Ω_2}, $L_{B_{02}}$, $L_{B_{20}}$ and L_{Ω_1} are given by eqs. (3.43)–(3.45) and (3.48) respectively. Here $\boldsymbol{\theta}$ are dependence parameters and $\boldsymbol{\sigma} = (\sigma_1, \sigma_2)$, $\boldsymbol{\xi} = (\xi_1, \xi_2)$, $\boldsymbol{\lambda} = (\lambda_1, \lambda_2)$ are the six marginal parameters. Since obtaining the contribution to the likelihood of points in the region Ω_1 requires the convergence of the iterative proportional fitting algorithm, evaluating the likelihood at a particular value of the parameters is computationally intensive. Also the fitting algorithm involves many one-dimensional integrals in order to obtain the margins of $g^{(k)}$.

3.6.3 *Non-Parametric Estimation*

Non-parametric estimation of multivariate extreme value distributions concerns estimation of the dependence measure H_* or its equivalent A, the dependence function.

The work started with Pickands (1981). If $\{(Y_{i,1}, Y_{i,2})\}$ come from (3.18) then he observes that $1/\max\{Y_{i,1}(1-w), Y_{i,2}w\}$ has an exponential distribution with mean $1/A(w)$, for each $w \in [0,1]$. Hence, with a sample $\{(Y_{i,1}, Y_{i,2})\}$, $1 \le i \le n$, an obvious consistent estimator of $A(w)$ is

$$A_n(w) = n\left[\sum_{i=1}^{n} 1/\max\{Y_{i,1}(1-w), Y_{i,2}w\}\right]^{-1}. \tag{3.49}$$

Besides being non-convex and non-differentiable, this estimator does not satisfy $A_n(0) = A_n(1) = 1$. However, the simple modification

$$\bar{A}_n(w) = n\left[\sum_{i=1}^{n} 1/\max\{Y_{i,1}(1-w), Y_{i,2}w\} - (1-w)\sum_{i=1}^{n} 1/Y_{i,2} - w\sum_{i=1}^{n} 1/Y_{i,1} + n\right]^{-1}$$

may be used to overcome this defect. Pickands proposes a method of making it also convex by replacing A_n by its convex minorant. But the construction is rather involved and leads to a function having a complicated implicit dependence upon the data. It is thus natural to seek other ways of estimating $A(.)$. Motivated by the fact that the sum of convex functions is convex, Tiago de Oliveira (1987) proposes the alternative estimate

$$A_{n,\delta_n}(w) = 1 - n^{-1} R(\delta_n) \sum_{i=1}^{n} \min\{Y_{i,1}^{\delta_n}(1-w), Y_{i,2}^{\delta_n} w\},$$

where $0 < \delta_n < 1$ is a sequence of exponents such that $\delta_n \uparrow 1$ as $n \to \infty$, and where $R(\delta)$ is a function of $0 < \delta < 1$ such that $R(\delta)/(1-\delta) \to 1$ as $\delta \uparrow 1$. Deheuvels and Tiago de Oliveira (1989) establish that this estimator is consistent for A if and only if the sequence $0 < \delta_n < 1$ satisfies the condition that $\delta_n \to 1$ and $(1-\delta_n)\log n \to \infty$ as $n \to \infty$.

Pickands did not even consider making (3.49) differentiable. Smith (1985) addresses this issue. He observes that (3.18) is differentiable if and only if h exists on $(0,1)$. Then $h(w) = A''(w)$. This suggests the estimator

$$h_n(w;\lambda) = \frac{A_n(w+\lambda) + A_n(w-\lambda) - 2A_n(w)}{\lambda^2}, \tag{3.50}$$

which is a very simple approximation to the second derivative of A. Here λ is an adjustable smoothing parameter. Asymptotic arguments in Smith *et al.* (1990) show that, for large n and small λ,

$$\begin{aligned} \text{Bias in } h_n(w;\lambda) &\sim \lambda^2 h''(w)/12, \\ \text{Variance of } h_n(w;\lambda) &\sim C(w)/(n\lambda), \end{aligned} \tag{3.51}$$

where

$$\begin{aligned} C(w) = \{&12A^2(w) + 12(1-2w)A(w)A'(w) - 12w(1-w)A'^2(w) \\ &+ 4w(1-w)A(w)A''(w)\}/\{3w^2(1-w)^2\}. \end{aligned} \tag{3.52}$$

The assumptions here are that $\lambda \le w \le 1-\lambda$, and that h is twice continuously differentiable at w. For $w < \lambda$, eq. (3.50) is no longer valid and we replace it by

$$h_n(w;\lambda) = \frac{2\{\lambda A_n(0) - (\lambda + w)A_n(w) + wA_n(\lambda + w)\}}{\lambda w(w+\lambda)}$$

with a similar modification for $w > 1-\lambda$.

The bias in (3.51) is of order λ^2, the variance is of order $(n\lambda)^{-1}$ and hence the minimum mean square error is $O(n^{-4/5})$ achieved when $\lambda = O(n^{-1/5})$. These are the same orders of magnitude as arise in ordinary one-dimensional density estimation (Silverman, 1986), under the assumption that h is twice differentiable.

Often a large value of λ (in range 0.25–0.5) is needed to obtain h_n positive for all w. This is because of the roughness of A_n and therefore suggests smoothing A_n before differentiating. The kernel method is widely used for smoothing problems in statistics (Silverman, 1986), so it is an obvious possibility to try here.

In its simplest form, the kernel estimator given in Smith *et al.* (1990) is

$$A_{n1}(w;\lambda) = \frac{1}{\lambda}\int_0^1 A_n(u)K\left(\frac{w-u}{\lambda}\right)du \qquad (0 < w < 1), \tag{3.53}$$

where λ is a smoothing parameter and K a kernel function, which in most applications is taken to be a pdf symmetric about 0. The integral in (3.53) is necessarily over a finite range, which creates difficulties when w is near 0 or 1. Differentiating (3.53) and assuming that K is twice differentiable, Smith *et al.* (1990) obtain estimators for A' and A''. Calculations similar to those leading (3.51) then give

$$\begin{aligned} \text{Bias in } A''_{n1}(w) &\sim \frac{\lambda^2 h''(w)}{2}\int_{-\infty}^{\infty} v^2 K(v)dv, \\ \text{Variance of } A''_{n1}(w) &\sim \frac{3C(w)}{2w\lambda}\int_{-\infty}^{\infty} K^2(v)dv, \end{aligned}$$

with $C(w)$ given by (3.52).

Calculations based on these approximations show that, using the criterion suggested by Silverman (1986, Sec. 3.3), the differencing method is only 1.4% less efficient than that based on the theoretically optimal Epachenikov kernel. Thus, from the point of view of mean square error, there is nothing in practice to choose between the two methods. Examples have shown that the kernel method yields smoother estimates, but behaves badly at the end points.

A second from of the kernel method suggested in Smith *et al.* (1990) is for the range of the integral to be first transformed from $(0,1)$ to $(-\infty,\infty)$. The transformation adopted is $w \to \log(w^{-1}-1)$ and the corresponding estimator is then

$$A_{n2}(w) = \frac{1}{\lambda}\int_{-\infty}^{\infty} A_n\left[\frac{1}{1+e^u}\right] K\left[\frac{\log(w^{-1}-1)-u}{\lambda}\right] du. \tag{3.54}$$

As before, Smith *et al.* (1990) estimate A' and A'' by direct differentiation in (3.54). The main practical advantage of (3.54) over (3.53) is that it works much better near the end points ($w = 0$ and 1).

A third procedure suggested in Smith *et al.* (1990) is a combination of the kernel method with Pickands' (1981) procedure: first define $A_n^c(.)$ to be the greatest convex minorant of $A_n(.)$, this is Pickands' estimator, and then apply (3.54) with A_n^c in place of A_n. No asymptotic results have been found for this procedure but it sometimes leads to more satisfactory results in practice.

Some recent work by Einmahl *et al.* (1993) constructs a non-parametric estimate of the dependence measure H_* for a random sample from the original distribution F. The construct for $p = 2$ is as follows.

Let $\{(X_{i,1}, X_{i,2})\}$ be a sequence of independent and positive random vectors with equal margins and common df $F \in D(G)$. Here G is a bivariate extreme value df of the form (3.8) with marginal dfs $\exp\{-x_j^{-\alpha}\}$, $x_j > 0$, $j = 1, 2$. Let $(\rho_i = \sqrt{X_{i,1}^2 + X_{i,2}^2}$, $\Theta_i = \arctan(X_{i,2}/X_{i,1}))$ be the pseudo-polar coordinates of $\{(X_{i,1}, X_{i,2})\}$. Let $\rho_{(k)}$ denote the kth order statistic from $\rho_1, \dots, \rho_n$ and k_n a sequence of positive integers satisfying

$$1 \le k_n \le n/2, \quad \text{and} \quad k_n \to \infty,\ k_n/n \to 0\ (n \to \infty). \tag{3.55}$$

Given k_n, the non-parametric estimate of $H_*([0, 1/(1+\tan\theta)]) = H_\dagger(\theta)$, say, is

$$H_n(\theta) = \frac{1}{k_n}\sum_{i=1}^{n} I\{\Theta_i \le \theta, \rho_i \ge \rho_{(n-k_n+1)}\}, \quad 0 \le \theta \le \pi/2 \tag{3.56}$$

which is both weakly and strongly consistent. Moreover, $k_n^{1/2}(H_n - H_\dagger)$ converges weakly to Λ, a mean zero Gaussian process with covariance structure

$$E\{\Lambda(\theta_1)\Lambda(\theta_2)\} = \Phi(\min(\theta_1,\theta_2)) - \Phi(\theta_1)\Phi(\theta_2), \quad 0 \le \theta_1, \theta_2 \le \pi/2.$$

Here $\Phi(.)$ denotes the cdf of the standard Normal distribution. Note also that $\Lambda =^d B \circ \Phi$, where B is a Brownian bridge.

We now outline an extension of the above result, due to Einmahl *et al.* (1997), for the case where $\{(X_{i,1}, X_{i,2})\}$ with joint df $F \in D(G)$ have unequal margins and G_j is Generalized

Extreme Value with parameters $(0,\ 1,\ \xi_j)$. Let $a_j(n) > 0$, $b_j(n) \in \Re$, $(j = 1, 2)$ be the norming constants for which

$$F^n(a_1(n)x_1 + b_1(n), a_2(n)x_2 + b_2(n)) \to G(x_1, x_2)$$

as $n \to \infty$. With k_n as defined in (3.55), suppose that $\tilde{a}_j(n/k_n)$, $\tilde{b}_j(n/k_n)$, $\tilde{\xi}_j$ are any estimators of $a_j(n/k_n)$, $b_j(n/k_n)$, ξ_j $(j = 1, 2)$ respectively satisfying,

$$\frac{\tilde{a}_j(n/k_n)}{a_j(n/k_n)} \to^P 1,\ \frac{\tilde{b}_j(n/k_n) - b_j(n/k_n)}{a_j(n/k_n)} \to^P 0,\ \tilde{\xi}_j \to^P \xi_j$$

as $n \to \infty$, $k_n \to \infty$ and $k_n/n \to 0$. For example, some specific estimators studied by Dekkers *et al.* (1989) satisfy these conditions. Finally, define the pseudo-polar coordinates as $\rho_i = \max(Y_{i,1}(n/k_n), Y_{i,2}(n/k_n))$ and $\Theta_i = \arctan(Y_{i,2}(n/k_n)/Y_{i,1}(n/k_n))$ where

$$Y_{i,j}(n/k_n) = \left[1 + \tilde{\xi}_j \frac{X_{i,j} - \tilde{b}_j(n/k_n)}{\tilde{a}_j(n/k_n)}\right]^{1/\hat{\xi}_j}, \qquad j = 1, 2.$$

Then the extension of (3.56) is the estimate:

$$H_n(\theta) = \frac{1}{k_n} \sum_{i=1}^{n} I\{\Theta_i \le \theta, \rho_i > 1\}.$$

This is weakly consistent for $H_\dagger(\theta)$ and is also strongly consistent under some further mild conditions. It is also asymptotically normal. But the expression for the associated covariance structure is too complicated.

The most recent non-parametric estimates are given in Caperaa *et al.* (1997). The estimates are based on the following representation.

Let (Y_1, Y_2) be distributed according to (3.18). Let $U_1 = G_{*1}(Y_1)$, $U_2 = G_{*2}(Y_2)$ and $Z = \log(U_1)/\log(U_1U_2)$. Then the df of the random variable Z is given by

$$P(z) = \Pr(Z \le z) = z + z(1 - z)D(z),$$

where $D(z) = A'(z)/A(z)$ and $A'(z)$ denotes the right derivative of A for all $0 \le z < 1$. A proof of this representation can be found in Ghoudi *et al.* (1998). As a consequence of this representation, one gets

$$\frac{A(t)}{A(s)} = \exp\left\{\int_s^t \frac{P(z) - z}{z(1 - z)} dz\right\}$$

for arbitrary choices of $0 \le s \le t \le 1$. Since $A(0) = A(1) = 1$, one may write

$$A(t) = \exp\left\{\int_0^t \frac{P(z) - z}{z(1 - z)} dz\right\} = \exp\left\{-\int_t^1 \frac{P(z) - z}{z(1 - z)} dz\right\}.$$

Let P_n be the empirical df of $Z_1, \ldots, Z_n$ for a random sample $\{(Y_{i,1},\ Y_{i,2}),\ i = 1, \ldots, n\}$ drawn from (3.18). Replacing P by P_n in the above expressions yields two possible non-parametric estimators for A, denoted by

$$A_n^0(t) = \exp\left\{\int_0^t \frac{P_n(z) - z}{z(1 - z)} dz\right\}, \quad A_n^1(t) = \exp\left\{-\int_t^1 \frac{P_n(z) - z}{z(1 - z)} dz\right\}.$$

Caperaa *et al.* (1997) show that $\log A_n^0$ is an unbiased and uniformly, strongly consistent estimator of $\log A$, i.e. $E(\log A_n^0) = \log A$ for all $n \geq 1$ and

$$\sup_{t\in[0,1]} |\log A_n^0(t) - \log A(t)| \to 0$$

almost surely. In addition, the process $n^{1/2}(\log A_n^0 - \log A)$ is asymptotically normal with zero mean and covariance matrix

$$\Gamma_0(s,t) = \int_0^s \int_0^t \frac{P(\min(u,v)) - P(u)P(v)}{uv(1-u)(1-v)} dvdu.$$

Since $P(\min(u,v)) \geq P(u)P(v)$, it should be observed that $\Gamma_0(t,t)$ is monotone increasing in t, so that, in spite of its attractive properties, $\log A_n^0(t)$ is an increasingly unreliable estimator of $\log A(t)$ as $t \to 1$. A similar analysis by Caperaa *et al.* (1997) shows that $n^{1/2}(\log A_n^1 - \log A)$ is asymptotically normal with zero mean and covariance matrix

$$\Gamma_1(s,t) = \int_s^1 \int_t^1 \frac{P(\min(u,v)) - P(u)P(v)}{uv(1-u)(1-v)} dvdu,$$

and hence that the variance of $\log A_n^1(t)$ is a decreasing function of t. This phenomenon suggests that a combined estimator of the following form might be preferable to each of the $\log A_n^i$:

$$\log A_n(t) = p(t) \log A_n^0(t) + \{1 - p(t)\} \log A_n^1(t),$$

where $p(t)$ is a bounded weight function on $[0,1]$ that gives comparatively more weight to $\log A_n^i$ in the neighbourhood of i. Again, Caperaa *et al.* (1997) show that $\log A_n$ is unbiased and uniformly, strongly consistent estimator of $\log A$. In addition, the process $n^{1/2}(\log A_n - \log A)$ is asymptotically normal with zero mean and variance function

$$\Gamma(t) = p^2(t)\Gamma_0(t,t) + \{1 - p(t)\}^2\Gamma_1(t,t) + 2p(t)\{1 - p(t)\}C(t),$$

where

$$C(t) = -\int_0^t \int_t^1 \frac{P(u)\{1 - P(v)\}}{uv(1-u)(1-v)} dvdu \leq 0$$

is the asymptotic covariance of $n^{1/2} \log A_n^0(t)$ and $n^{1/2} \log A_n^1(t)$. The choice of $p(t)$ that minimises the asymptotic variance, $\Gamma(t)$, is:

$$p(t) = \frac{\Gamma_1(t,t) - C(t)}{\Gamma_0(t,t) + \Gamma_1(t,t) - 2C(t)}.$$

We now define the new estimator in terms of A_n. If $Z_{(1)}, \ldots, Z_{(n)}$ stand for the ordered Z_i's, and if

$$Q_i = \left\{ \prod_{k=1}^{i} \frac{Z_{(k)}}{(1 - Z_{(k)})} \right\}^{1/n}, \quad 1 \leq i \leq n$$

then Caperaa *et al.* (1997) note that A_n can be written in closed form as

$$A_n(t) = \begin{cases} (1-t)Q_n^{1-p(t)} & \text{if } 0 \le t \le Z_{(1)}, \\ t^{i/n}(1-t)^{1-i/n}Q_n^{1-p(t)}Q_i^{-1} & \text{if } Z_{(i)} \le t \le Z_{(i+1)} \ (1 \le i \le n-1), \\ tQ_n^{-p(t)} & \text{if } Z_{(n)} \le t \le 1, \end{cases}$$

provided that the Z_i's are distinct. This estimator satisfies $A_n(0) = A_n(1) = 1$, provided that $p(0) = 1 - p(1) = 1$. It is an asymptotically unbiased estimator of A and is also uniformly, strongly consistent.

3.7 Simulation

This section is concerned with methodologies for simulation of bivariate extremes. We discuss, in turn, three known approaches due to Shi *et al.* (1993), Ghoudi *et al.* (1998) and Nadarajah (1999b).

Shi *et al.* (1993) describe a scheme for simulating (Y_1, Y_2) from the bivariate symmetric logistic distribution, (3.21). Defining the transformations $1/Y_1 = Z\cos^{2/q}V$ and $1/Y_2 = Z\sin^{2/q}V$, they note that the joint density of (Z, V) factorizes as

$$(q^{-1}z + 1 - q^{-1})e^{-z}\sin 2v, \ 0 < v < \pi/2, \qquad 0 < z < \infty$$

which shows that Z and V are independent. It is easily characterized that V may be represented as $(\arcsin U^{1/2})$, where U is uniform on $(0, 1)$, while Z is the $1-q^{-1} : q^{-1}$ mixture of a unit exponential random variable and the sum of two independent unit exponential random variables. Hence this suggests an easy way of simulating from (3.21).

Ghoudi *et al.* (1998) describe a scheme that is applicable for all bivariate extreme value distributions. To begin with note that the copula, i.e. the df with uniform margins, associated with (3.18) is

$$D(u_1, u_2) = \exp\left[(\log u_1 + \log u_2)A\left(\frac{\log u_1}{\log u_1 + \log u_2}\right)\right], \quad 0 < u_1, u_2 < 1.$$

Ghoudi *et al.* show that the joint df of $Z = Y_2/(Y_1 + Y_2)$ and $V = D(\exp(-1/Y_1), \exp(-1/Y_2))$ is

$$G_{Z,V}(z, v) = v\int_0^z \frac{t(1-t)}{A(t)}dA'(t) + (v - v\log v)\left\{z + z(1-z)\frac{A'(z)}{A(z)} - \int_0^z \frac{t(1-t)}{A(t)}dA'(t)\right\}.$$

It follows from this that the marginal df of Z is

$$G_Z(z) = z + z(1-z)\frac{A'(z)}{A(z)} \tag{3.57}$$

and that the conditional df of $V|Z = z$ is

$$\frac{1}{g_Z(z)}\frac{\partial G_{Z,V}(z, v)}{\partial z} = vp(z) + (v - v\log v)\{1 - p(z)\}$$

where

$$p(z) = \frac{z(1-z)A''(z)}{A(z)g_Z(z)}$$

and g_Z is the derivative of G_Z. Thus, given Z, the law of V is uniform on $[0, 1]$ with probability $p(Z)$ and equal to the law of the product of two independent uniforms on $[0, 1]$ with probability $1 - p(Z)$. Hence, to simulate (Y_1, Y_2) from (3.18), one can use the following procedure:

- simulate Z according to the distribution given by (3.57);
- Having Z, take $V = U_1$ with probability $p(Z)$ and $V = U_1U_2$ with probability $1-p(Z)$. Here U_1 and U_2 are independent uniforms on $[0, 1]$;
- Set $Y_1 = V^{Z/A(Z)}$ and $Y_2 = V^{(1-Z)/A(Z)}$.

The most recent method is due to Nadarajah (1999b). It differs from the two schemes above in that it does not simply simulate from a bivariate extreme value distribution, but uses the limiting point process result, (3.5), as an approximation to simulate bivariate extreme values. In (3.5), we assume, for large n, that $\mathcal{P}_n$ coincides with the Poisson process with intensity satisfying eq. (3.6) for a region $B \subset \{[0, \infty) \times [0, \infty)\} \backslash \{(0, 0)\}$ sufficiently away from $\{(0, 0)\}$. Thus, simulation of (Y_1, Y_2) in B, under this model, reduces to simulation of a Poisson process restricted to B. There are standard procedures to simulate from a Poisson process, we consider a simple one. Take $B = B_0 = \{(y_1, y_2) : y_1 + y_2 > r_0, y_1 \geq 0, y_2 \geq 0\}$ with r_0 sufficiently large so that the assumption is valid; see below for simulation over general forms of B. The conditional pdf of $(R, W) = T(Y_1, Y_2) = (Y_1 + Y_2, Y_1/(Y_1 + Y_2))$ over $T(B_0)$ is:

$$r^{-2}H_*(dw) \Big/ \int_{r_0}^{\infty} \int_0^1 s^{-2} ds H_*(dv) = r_0 r^{-2} 2^{-1} H_*(dw),$$

using eq. (3.6). Since this conditional density factorizes, i.e. R and W are independent, we can simulate the radial and angular coordinates independently of each other. The procedures for these are as follows:

- We simulate r by the inversion principle: set $\int_{r_0}^{r} r_0 s^{-2} ds = u$ for $u \sim U(0, 1)$ and invert to obtain $r = r_0/(1 - u)$;
- Since H_* is a composition of the density h in the interior, $(0, 1)$, and the atoms, $H_*(\{0\})$ and $H_*(\{1\})$, at the end points, simulation of w can be performed by the method of composition (see e.g. Ripley, 1987, Sec. 3.2) in two steps. Firstly, set w as 0, belonging to $(0, 1)$ or 1 with probabilities $2^{-1}H_*(\{0\})$, $1 - 2^{-1}H_*(\{0\}) - 2^{-1}H_*(\{1\})$ and $2^{-1}H_*(\{1\})$, respectively. Then, if w is set to be in $(0, 1)$, simulate its specific value from the pdf

$$h^*(w) = \frac{h(w)}{2 - H_*(\{0\}) - H_*(\{1\})}$$

using the rejection method (see e.g. Ripley, 1987, Sec. 3.2). To apply the rejection method to simulate from $h^*(w)$ we need to determine a probability density $g(w)$, from which it is easy to simulate, and a constant M such that $h^*(w)/g(w) \leq M$ for all

$w \in (0,1)$. The density g is referred to as the envelope. Under mild conditions on h^*, a form for the envelope g is easily established by the following result: if h^* is a continuous probability density over $(0,1)$ and there are some constants $q_0 > -1$, $q_1 > -1$ such that $h^*(w) = O(w^{q_0})$ and $h^*(1-w) = O(w^{q_1})$ hold as $w \to 0$ then the beta density

$$g(w) = \frac{w^{q_0}(1-w)^{q_1}}{Be(q_0+1, q_1+1)}, \quad w \in (0,1)$$

where

$$Be(a,b) = \frac{\Gamma(a)\Gamma(b)}{\Gamma(a+b)}, \quad a > 0, b > 0$$

satisfies $h^*(w)/g(w) \leq M < \infty$ for some constant M. Nadarajah (1994) shows that each currently known model for H_* (including those discussed in Sec. 3.4) satisfies the conditions of this result for particular choices of q_0, q_1 and M. Hence, using the rejection method of simulation for w reduces, for each model, to simulation from a Beta(q_0+1, q_1+1) distribution for which routines are widely available.

To simulate (Y_1, Y_2) over a region B not having the form B_0 we follow the usual practice: first apply these procedures to simulate (Y_1, Y_2) over an B_0 for which $B \subset B_0$ and then delete those points falling outside of B.

3.8 A Selective Survey of Applications of Multivariate Extreme Value Distributions

This is the third separate section in this book devoted to applications of extreme value distributions. The rationale behind having individual sections on applications of classical univariate extreme value distributions, generalized (univariate) extreme value distributions and now on multivariate ones is to minimize the overall pressure on the reader and reduce the amount of information to be absorbed in a single setting. Also since most of the readers will, hopefully, be studying this monograph in an organized manner, it is indeed more convenient to have the section of the theory of multivariate extreme value distributions to be immediately supplemented by the examples of practical applications of this theory. Indeed — as in the previous applications sections — we shall provide only examples rather than attempting a complete survey which is beyond the scope of the book. Again recent contributions by J. Tawn and his associates will receive special attention since — in the opinion of the authors — this work contains laudable achievements and examples of combining important theoretical results with efficient utilization of modern computer technology and recent inference procedures based on it.

3.8.1 *Some Earlier Applications*

In his survey of applications of multivariate extremes, Tawn (1994) points out that many problems that involve applications of extreme value methodology are "inherently multivariate by nature". Indeed, observations of a number of different physical processes observed at one site, or a number of summarizing features of behavior of a single process at a particular location, or consecutive observations during extreme events of one process, or most prominently a spatial process observed at a finite number of sites are all examples of observations leading naturally to multivariate extreme value distributions.

First, however, we shall comment on an earlier contribution by Gumbel and Goldstein (1964) involving the most popular, almost classical, topic of extreme value theory: ages at death classified according to sex. The well-known fact that the oldest ages of women are higher than those for men, requires testing of the natural hypothesis of independence of oldest ages at death. The authors, using the data of the oldest ages at death for men (x_1) and for women (x_2) in Sweden for the period 1905–58 ($n = 54$), perform a classical χ^2-test of independence. Let y_j denote the transform of x_j to the Gumbel margin ($j = 1, 2$). Under the hypothesis of independence, the bivariate asymptotic df of (y_1, y_2) is

$$G(y_1, y_2) = \exp(-e^{-y_1} - e^{-y_2}).$$

To apply the χ^2-test, the authors enumerate the number of data in the five intervals, given by $G(y_1, y_2) < 0.2; \Delta G(y_1, y_2) = 0.2; G(y_1, y_2) > 0.8$. The expected number in each interval is $54/5 = 10.8$. The observed numbers for the oldest ages are given in the table below.

Interval	Observed Number of Oldest Ages
$0.0 \le G(y_1, y_2) \le 0.2$	11
$0.2 < G(y_1, y_2) \le 0.4$	12.5
$0.4 < G(y_1, y_2) \le 0.6$	7
$0.6 < G(y_1, y_2) \le 0.8$	9.5
$0.8 < G(y_1, y_2) \le 1.0$	14
Total	54
χ^2	2.266
p-value	0.69

The number of degrees of freedom is four which leads to the p-value given in the last line. Thus, the conclusion is that the hypothesis of independence should be accepted.

We extended the above analysis to the years 1967–97 (thanks to Statistics Sweden and to Dr. Eva Elver for providing the relevant data). The new data showed an increasing gap between the ages of men and women and a somewhat erratic slow increase in the values of the maxima. But the hypothesis of independence was found to be valid for the new and combined data.

Gumbel and Goldstein (1964) also carried out a similar analysis to floods data from the Ocmulgee river in the State of Georgia, U.S.A. The data were annual maximum discharges and came from two different stations: one upstream (x_1) and the other downstream (x_2). The analysis (again involving the χ^2-test) leads to rejection of the assumption of independence between the upstream and downstream floods.

As pointed out by Tawn (1994), the paper by Gumbel and Goldstein (1964) reviewed above and the subsequent one by Gumbel and Mustafi (1967) in which annual maximum river flows of the Fox river at two sites in the State of Wisconsin (U.S.A.) were studied seem to be the only papers in sixties which demonstrated applicability of the theory of bivariate extremes. To the best of our knowledge, there are indeed no descriptions of work dealing with application of the theory of bivariate or multivariate extreme values (with a possible exception of McFadden (1978)) until mid eighties. After that we encounter a flood of papers on multivariate applications by de Haan, Smith, Tawn and their associates. This burgeoning activity is the result of advances in probabilistic results (see e.g. de Haan (1985)) and new approaches to estimation of the dependence structure of multivariate extremes (see e.g. Smith (1994)) the latter being the main aspect of multivariate extreme value methodology.

3.8.2 *Directional Modelling of Wind Speeds*

Coles and Walshaw (1994) investigate directional modelling of wind speeds based on data collected at the University of Sheffield for the U.K. Meteorological Office. The data consist of hourly maximum wind gust speeds recorded in knots, together with the gust direction, recorded to the nearest 10°.

The authors emphasize the importance of dependence of extreme winds across directions because extreme storms tend to change direction providing successive extreme values in several directions. They suggest the following model for the joint distribution of annual maximum wind speeds for any set of directions $\Theta \subset (0, 2\pi]$:

$$\Pr\{X_\theta \leq x_\theta, \quad \forall\, \theta \in \Theta\} = \exp\left[-\int_0^{2\pi} \max_{\theta\in\Theta}\left\{\frac{f_0(\omega;\theta,\eta)}{y_\theta}\right\} d\omega\right],$$

where

$$y_\theta = \{1 + \xi_\theta(x_\theta - \mu_\theta)/\sigma_\theta)\}^{1/\xi_\theta}$$

and f_0 is the von Mises circular density. This is the so-called circular max-stable model. A special case of this model is the bivariate circular extreme value distribution discussed in Sec. 3.4.6. The model can be used to calculate probabilities of joint events across entire sectors.

For the Sheffield data the authors model directions within each of the four quadrants

$$(0°, 90°], (90°, 180°], (180°, 270°] \text{ and } (270°, 360°]$$

separately. They find some indication that the mechanisms of dependence in quadrants 1 and 3 are different, dependence being generally weaker in quadrant 1 than in quadrant 3. This may have a physical explanation. The figure below graphs the return level curves for each quadrant as a function of plotting quantiles

$$p = 1 - \Pr\{X_\theta \leq x, \quad \forall\theta \in \Theta\}$$

on the scale $-\log(-\log(1-p))$. The substantially different extremal behavior within each quadrant demonstrated by these plots is the kind of information which could usefully be applied in engineering design.

3.8.3 *Applications to Structural Design*

As it was no doubt observed from previous chapters, one of the aims of extreme value analysis is to model the extreme tail to estimate probabilities of failure of some structure. Coles and

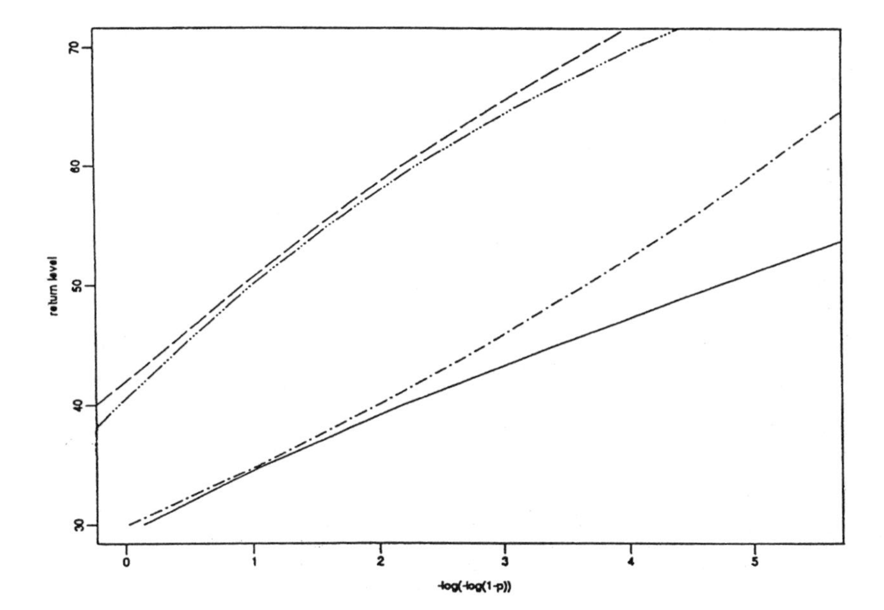

Figure 3.11: Annual maximum return levels, for exceedance probability $1 - p$, for wind speeds within each quadrant:__________, quadrant 1; ._____ ._____, quadrant 2; _____ _____, quadrant 3; _____ . . ._____, quadrant 4.

Tawn (1994) assert that many forms of structure fail owing to a combination of various processes at extreme levels. Thus, multivariate extreme value methodology is an adequate tool for quantifying the risk of failure due to extreme levels.

Let $X_1, \ldots, X_p$ be random variables corresponding to the constituent processes that affect some structure. Coles and Tawn (1994) express the concept of failure in terms of the value which is exceeded by an appropriately chosen function of the univariate random variables. Postulating the existence of structural parameter ν, it is assumed that structural failure occurs if

$$\mathbf{X} = (X_1, \ldots, X_p)$$

belongs to some failure region B_ν in $\Re^p$. Explicitly,

$$B_\nu = \{\mathbf{x} \in \Re^p : b(\mathbf{x}, \nu) > 0\}$$

for some "boundary function" $b : \Re^p \times \mathcal{V} \to \Re$ where $\mathcal{V}$ is the parameter space for ν. For a given ν it is required to determine

$$p = \Pr(\mathbf{X} \in B_\nu)$$

or conversely to determine a value of ν which gives rise to a specified value of p. Imposing further optimization criteria can lead to unique solutions for ν.

Numerous practical problems relating to extremes especially in environmental setting can be stated in the above manner. Coles and Tawn (1994) cite examples in offshore engineering (e.g. oil-rigs), in assessment of extremal total volume of rainfall from a discrete set of sites and in structural failures of river banks (where the corresponding failure region has a componentwise structure). In the rest of section we provide some details of two far-reaching examples studied in this context.

The first example concerns reservoir flood safety (Anderson and Nadarajah, 1993; Anderson *et al.*, 1994). It arises from a project sponsored by the U.K. Department of Environment involving researchers from the University of Sheffield and the Institute of Hydrology at Wallingford with the aim to provide a basis for the reassessment and modification of current British design recommendations for reservoirs.

If a reservoir receives a sudden inflow of water during a very violent storm then it is conceivable that the provision for outflow at the reservoir's dam could prove inadequate, resulting in an unplanned overflow. Such overflows could cause serious damage to the dam structure, and it could possibly lead to catastrophic failure, with serious economic and environmental consequences, and perhaps even loss of life. Using the above terminology, a plausible boundary function is

$$b(R, W, S; \nu) = \Delta(R, W, S; \nu) - l$$

where

- R and S are total rainfall and total snowmelt during the storm over the area draining into the reservoir;
- W is the wind speed at some time after the onset of the storm when the inflow to the reservoir has had time to raise the water level;
- Δ is a function that computes the peak water level at the dam wall during the storm. It will be a complicated deterministic function, depending on specific characteristics such as reservoir size, drawoff rate and orientation;
- and, l is the height of the dam wall.

The probabilities, p, of structural failure are computed by modelling the joint extremes of (R, W, S) by a trivariate extreme value distribution and thus inducing a probability distribution on $\Delta(R, W, S)$. Figures 3.12 and 3.13 show typical probabilities associated with rare water levels for a number of different settings of the reservoir characteristics. The probabilities are expressed in terms of return periods, the mean interval (in years) between exceedances of the levels.

The second example concerns the safety of sea dikes in the Netherlands (Brunn and Tawn, 1998; de Haan and de Ronde, 1998). It arises from the European Union undertaking "Neptune" which involved three institutions from the U.K. (British Maritime Technology, Lancaster University and the University of East Anglia), three institutions from the Netherlands (National Institute for Coastal and Marine Management, Delft Hydraulics and Erasmus University Rotterdam) and the GKSS Forchungszentrum Geesthacht in Germany. Needless to say that the project has a substantial practical value for the Netherlands (40% of the country is below the mean sea level and is protected by dikes). In 1953, the sea dikes broke in part of the country and the subsequent flooding claimed nearly 2000 lifes. The aim of the "Neptune" project was to estimate the probability of failure of a dike called "Pettemer zeedijk" which protects a gap in the natural coast protection formed by sand dunes near the

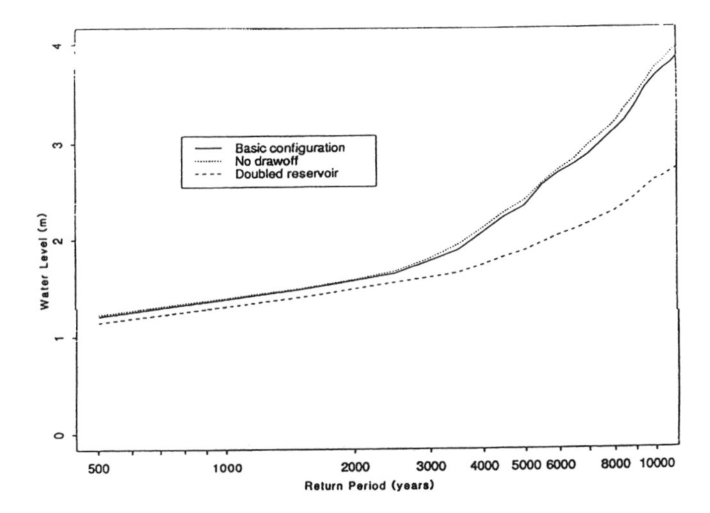

Figure 3.12: Effects of drawoff rate and reservoir size on probabilities of rare water levels.

town of Petten. The failure was taken to occur if 2% runup level exceeded the crest height of the dike. This 2% runup level is a function of the onshore sea state, involving variables such as wave height (HmO), wave direction (θ) and still water level (SWL). To simplify analysis wave direction, θ, was taken fixed perpendicular to the dike. A possible boundary function is that

$$b(HmO, SWL) = 0.3HmO + SWL - 7.6.$$

The estimation of p entails modelling of the joint tail of (HmO, SWL) using bivariate extreme value distributions. The work by the Dutch team, L. de Haan and J. de Ronde, gave the estimate

$$\hat{p} = 1.14 \times 10^{-3}$$

with the 95% confidence interval

$$(0, 8.75 \times 10^{-3}).$$

They noted also that zero probability is a real possibility. The results of investigation (on the same problem) by the British team, J. Brunn and J. Tawn, can be summarized as follows:

- the 100-year return level of HmO is 7.62m with 95% confidence interval (6.34, 9.03);
- the 100-year return level of SWL is 2.88m with 95% confidence interval (2.42, 3.47);
- the 100-year return level for the *structure variable*, 0.3 $HmO + SWL$, is 5.17m. Since the height of the sea dike under consideration is 7.6m high this estimate suggests that flooding is not likely (consistent with the result noted by the Dutch team). In fact, the return period associated with 7.6m is estimated to be 10,800 years.

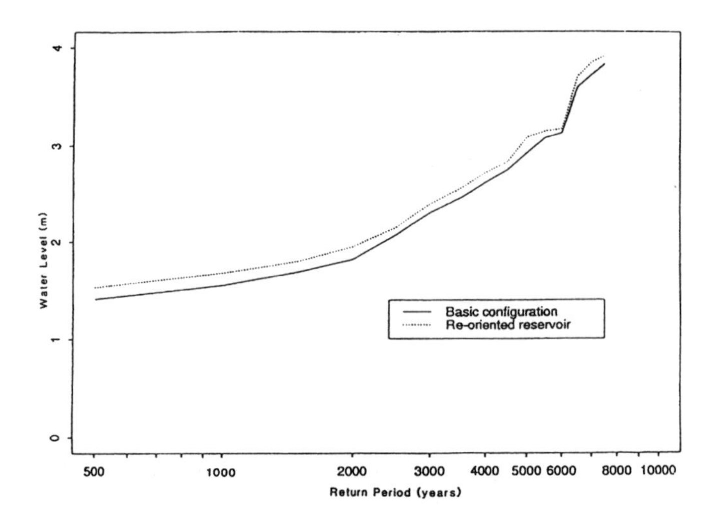

Figure 3.13: Effects of orientation on probabilities of rare water levels.

Other applications of multivariate extreme values (not discussed above) include: study of the dependence between the extreme concentrations of a pollutant at several monitoring stations in a region (Joe *et al.*, 1992; Joe, 1994); modelling of periods of extremely cold temperatures (Coles *et al.*, 1994; Bortot and Tawn, 1998); estimation of extreme sea levels in coastal engineering (Dixon and Tawn, 1994); estimation of the extremal behavior of the rainfall regime within a specified area, such as a catchment region, using pointwise measurements of rainfall from a network of sites (Coles and Tawn, 1996a); estimation of the probability of coastal flooding at an existing flood defence structure and for aiding the design of a new structure (Morton and Bowers, 1996; Bruun and Tawn, 1998; Zachary *et al.*, 1998); flood frequency analysis (EscalanteSandoval, 1998); and, estimation of rainfall-depth-frequency curves (Nadarajah *et al.*, 1998).

Bibliography

Abdelhafez, M. E. M. and Thomas, D. R. (1990). Approximate prediction limits for the Weibull and extreme value regression models, *Egyptian Statist. J.* **34**, 408–419.

Abdelhafez, M. E. M. and Thomas, D. R. (1991). Bootstrap confidence bands for the Weibull and extreme value regression models with randomly censored data, *Egyptian Statist. J.* **35**, 95–109.

Abramowitz, M. and Stegun, I. A. (eds.) (1964). *Handbook of Mathematical Functions with Formulas, Graphs and Mathematical Tables*, Dover.

Achcar, J. A. (1991). A useful reparametrization for the extreme value distribution, *Comput. Statist. Quart.* **6**, 113–125.

Achcar, J. A., Bolfarine, H. and Pericchi, L. R. (1987). Transformation of survival data to an extreme value distribution, *Statistician* **36**, 229–234.

Ahmad, M. I., Sinclair, C. D. and Spurr, B. D. (1988). Assessment of flood frequency models using empirical distribution function statistics, *Water Resources Res.* **24**, 1323–1328.

Ahmed, E. (1989). On the probability of selecting extreme value populations with the smallest location parameters, *Statistician* **38**, 191–195.

Ahsanullah, M. (1990). Estimation of the parameters of the Gumbel distribution based on the m record values, *Comput. Statist. Quart.* **5**, 231–239.

Ahsanullah, M. (1991). Inference and prediction problems of the Gumbel distribution based on record values, *Pakistan J. Statist.* B, **7**, 53–62.

Ahsanullah, M. (1994). Records of the generalized extreme value distribution, *Pakistan J. Statist.* A, **10**, 147–170.

Ahsanullah, M. and Holland, B. (1994). On the use of record values to estimate the location and scale parameters of the generalized extreme value distribution, *Sankhyā*, B, to appear.

Aitkin, M. and Clayton, D. (1980). The fitting of exponential, Weibull and extreme value distributions to complex censored survival data using GLIM, *Appl. Statist.* **29**, 156–163.

Al-Abbasi, J. N. and Fahmi, K. J. (1991). GEMPAK: a Fortran-77 program for calculating Gumbel's first, third, and mixture upper earthquake magnitude distribution employing maximum likelihood estimation, *Comput. Geosci.* **17**, 271–290.

Alves, M. I. F. and Gomes, M. I. (1996). Statistical choice of extreme value domains of attraction — a comparative analysis, *Comm. Statist. Theory and Methods* **25**, 789–811.

Aly, E.-E. A. A. and Shayib, M. A. (1992). On some goodness-of-fit tests for the normal, logistic and extreme-value distributions, *Comm. Statist. Theory and Methods* **21**, 1297–1308.

Anderson, C. W. (1970). Extreme value theory for a class of discrete distributions with applications to some stochastic processes, *J. Appl. Prob.* **7**, 99–113.

Anderson, C. W. and Coles, S. G. (1997). Approximations for Poisson-like maxima, *Ann. Appl. Probab.* **7**, 953–971.

Anderson, C. W., Dwyer, I. J., Nadarajah, S., Reed, D. W. and Tawn, J. A. (1994). Maximum reservoir water levels, in *Reservoir Safety and Environment*, 200–213. Thomas Telford.

Anderson, C. W. and Nadarajah, S. (1993). Environmental factors affecting reservoir safety, in *Statistics for the Environment I*, 163–182, eds. V. Barnett and K. F. Turkman, John Wiley and Sons.

Angus, J. E. (1993). Asymptotic theory for bootstrapping the extremes, *Comm. Statist. Theory and Methods* **22**, 15–30.

Arnold, B. C., Balakrishnan, N. and Nagaraja, H. N. (1992). *A First Course in Order Statistics*, John Wiley and Sons.

Ashour, S. K. and El-Adl, Y. M. (1980). Bayesian estimation of the parameters of the extreme value distribution, *Egyptian Statist. J.* **24**, 140–152.

Aziz, P. M. (1955), (1956). Application of the statistical theory of extreme values to the analysis of maximum pit depth data for aluminum, *Corrosion* **12**, 35–46. See also *ibid*, 495–506.

Bai, J., Jakeman, A. J. and McAleer, M. (1992). On the use of extreme value distributions for predicting the upper percentiles of environmental quality data, *Math. Comput. Simulation* **33**, 483–488.

Bain, L. J. (1972). Inferences based on censored sampling from the Weibull or extreme-value distribution, *Technometrics* **14**, 693–702.

Balakrishnan, N., Ahsanullah, M. and Chan, P. S. (1992). Relations for single and product moments of record values from Gumbel distribution, *Statist. Probab. Lett.* **15**, 223–227.

Balakrishnan, N. and Chan, P. S. (1992a). Order statistics from extreme value distribution, I: Tables of means, variances and covariances, *Comm. Statist. Simulation Comput.* **21**, 1199–1217.

Balakrishnan, N. and Chan, P. S. (1992b). Order statistics from extreme value distribution, II: Best linear unbiased estimates and some other uses, *Comm. Statist. Simulation Comput.* **21**, 1219–1246.

Balakrishnan, N. and Chan, P. S. (1992c). Extended tables of means, variances and covariances of order statistics from the extreme value distribution for sample sizes up to 30. *Report*, Department of Mathematics and Statistics, McMaster University, Hamilton, Canada.

Balakrishnan, N. and Chan, P. S. (1992d). Extended tables of best linear unbiased estimates from complete and type-II censored samples from the extreme value distribution for sample sizes up to 30. *Report*, Department of Mathematics and Statistics, McMaster University, Hamilton, Canada.

Balakrishnan, N., Chan, P. S. and Ahsanullah, M. (1993). Recurrence relations for moments of record values from generalized extreme value distribution, *Comm. Statist. Theory and Methods* **22**, 1471–1482.

Balakrishnan, N. and Cohen, A. C. (1991). *Order Statistics and Inference: Estimation Methods*, Academic Press.

Balakrishnan, N., Gupta, S. S. and Panchapakesan, S. (1992). Estimation of the location and scale parameters of the extreme value distribution based on multiply type-II censored samples. *Technical Report*, Department of Statistics, Purdue University.

Balakrishnan, N. and Leung, M. Y. (1988). Order statistics from the type I generalized logistic distribution, *Comm. Statist. Simulation Comput.* **17**, 25–50.

Balakrishnan, N. and Varadan, J. (1991). Approximate MLEs for the location and scale parameters of the extreme value distribution with censoring, *IEEE Trans. Reliab.* **40**, 146–151.

Ballerini, R. (1987). Another characterization of the type-I extreme value distribution, *Statist. Probab. Lett.* **5**, 83–85.

Barakat, H. M. (1997). On the continuation of the limit distributions of the extreme and central terms of a sample, *Test* **6**, 351–368.

Barakat, H. M. (1998). Weak limit of the sample extremal quotient, *Austral. New Zealand J. Statist.* **40**, 1, 83–93.

Barndorff-Nielsen, O. (1964). On the limit distribution of the maximum of a random number of independent random variables, *Acta. Math. Acad. Hungar.* **15**, 399–403.

Barnett, V. (1999). Ranked set sample design for environmental investigations, *Environmental Ecolog. Statist.* **6**, 59–74.

Barnett, V. and Moore, K. L. (1997). Best linear unbiased estimates in ranked set sampling with particular reference to imperfect ordering, *J. Appl. Statist.* **24**, 699–710.

Beirlant, J. and Devroye, L. (1999). On impossibility of estimating densities in the extreme tail, *Statist. Probab. Lett.* **43**, 57–64.

Beirlant, J., de Waal, D. J. and J. L. Teugels (1999). The generalized Burr–Gamma family of distributions with applications in extreme value analysis, Technical Report 262: Dept. of Mathematical Statistics, University of the Orange Free State, South Africa.

Beirlant, J. and Teugels, J. L. (1989). Asymptotic normality of Hill's estimator, in *Lecture Notes in Statistics*, **51**, 148–155, eds. J. Hüsler and R.-D. Reiss, Springer-Verlag.

Beirlant, J., Teugels, J. L. and Vynckier, P. (1996). *Practical Analysis of Extreme Values*, Leuven University Press.

Beirlant, J., Vynckier, P. and Teugels, J. L. (1996). Tail index estimation, Pareto quantile plots and regression diagnostics, *J. Amer. Statist. Assoc.* **91**, 1659–1667.

Beran, M., Hosking, J. R. M. and Arnell, N. (1986). Comment on "Two-component extreme value distribution for flood frequency analysis" by Fabio Rossi, Mauro Fiorentino, Pasquale Versace, *Water Resources Res.* **22**, 263–266.

Berger, J. O. (1980). *Statistical Decision Theory, Foundations, Concepts and Methods*, Springer-Verlag.

Berman, S. M. (1962). Limiting distribution of the maximum term in sequences of dependent random variables, *Ann. Math. Statist.* **33**, 894–908.

Berred, M. (1992). On record values and the exponent of a distribution with regular varying upper tail, *J. Appl. Probab.* **29**, 575–586.

Bhoj, D. S. (1997). Estimation of parameters of the extreme value distribution using ranked set sampling, *Comm. Statist. Theory and Methods* **26**, 653–667.

Billman, B. R., Antle, C. E. and Bain, L. J. (1972). Statistical inference from censored Weibull samples, *Technometrics* **14**, 831–840.

Bishop, Y. M., Fienberg, S. and Holland, P. (1975). *Discrete Multivariate Analysis*, MIT Press.

Blom, G. (1958). *Statistical Estimates and Transformed Beta-Variables*, Almquist and Wiksell.

Boos, D. D. (1984). Using extreme value theory to estimate large percentiles, *Technometrics* **26**, 33–39.

Borgman, L. E. (1961). The frequency distribution of near extreme, *J. Geophys. Res.* **66**, 3295–3307.

Bortkiewicz, L., von (1922). Variationsbreite und mittlerer Fehler, *Sitzungsber. Berli. Math. Ges.* **21**, 3–11.

Bortot, P. and Tawn, J. A. (1998). Models for the extremes of Markov chains, *Biometrika* **85**, 851–867.

Box, G. E. P. and Tiao, G. C. (1962). A further look on robustness via Bayes theorem, *Biometrika* **49**, 419–432.

Broussard, J. P. and Booth, G. G. (1998). The behavior of extreme values in Germany's stock index futures: An application to intradaily margin setting, *European J. Oper. Res.* **104**, 393–402.

Bruun, J. T. and Tawn, J. A. (1998). Comparison of approaches for estimating the probability of coastal flooding, *Appl. Statist.* **47**, 405–423.

Buishand, T. A. (1985). The effect of seasonal variation and serial correlation on the extreme value distribution of rainfall data, *J. Climate Appl. Meteor.* **25**, 154–160.

Buishand, T. A. (1989). Statistics of extremes in climatology, *Statist. Neerlandica* **43**, 1–30.

Burton, P. W. and Makropoulos, K. C. (1985). Seismic risk of circum-Pacific earthquakes: II. Extreme values using Gumbel's third distribution and the relationship with strain energy release, *Pure Appl. Geophys.* **123**, 849–866.

Campbell, J. W. and Tsokos, C. P. (1973). The asymptotic distribution of maxima in bivariate samples, *J. Amer. Statist. Assoc.* **68**, 734–739.

Canard, V. A. (1946). *Methods in Climatology*, Harvard University Press.

Canfield, R. V. (1975). The type-I extreme-value distribution in reliability, *IEEE Trans. Reliab.* **24**, 229–236.

Canfield, R. V. and Borgman, L. E. (1975). Some distributions of time to failure for reliability applications, *Technometrics* **17**, 263–268.

Caperaa, P., Fougeres, A. L. and Genest, C. (1997). A non-parametric estimation procedure for bivariate extreme value copulas, *Biometrika* **84**, 567–577.

Castillo, E. (1988). *Extreme Value Theory in Engineering*, Academic Press.

Chan, L. K., Cheng, S. W. H. and Mead, E. R. (1972). An optimum *t*-test for the scale parameter of an extreme-value distribution, *Naval Res. Logist. Quart.* **19**, 715–723.

Chan, L. K. and Kabir, A. B. M. L. (1969). Optimum quantiles for the linear estimation of the parameters of the extreme-value distribution in complete and censored samples, *Naval Res. Logist. Quart.* **16**, 381–404.

Chandra, M., Singpurwalla, N. D. and Stephens, M. A. (1981). Kolmogorov statistics for tests of fit for the extreme-value and Weibull distribution, *J. Amer. Statist. Assoc.* **76**, 729–731.

Changery, M. J. (1982). Historical extreme winds for the United States — Atlantic and Gulf of Mexico coastlines. U.S. Nuclear Regulatory Commission, NUREG/CR–2639.

Chao, A. and Hwang, S.-J. (1986). Comparison of confidence intervals for the parameters of the Weibull and extreme value distributions, *IEEE Trans. Reliab.* **35**, 111–113.

Chen, G. and Balakrishnan, N. (1995). The infeasibility of probability weighted moments estimation of some generalised distributions, in *Lifetime Reliability: Volume in Honor of A. C. Cohen*, 565–570, CRC Press.

Cheng, S. and Pan, J. (1995). Asymptotic expansions of estimators in extreme statistics, *Procs. of the 50th Session of the I.S.I.*, Beijing, IP15.1, 593–605.

Cheng, S., Peng, L. and Qi, Y. (1998). Almost sure convergence in extreme value theory, *Math. Nachr.* **190**, 43–50.

Chiou, P. (1988). Shrinkage estimation of scale parameter of the extreme-value distribution, *IEEE Trans. Reliab.* **37**, 370–374.

Chowdhury, J. U., Stedinger, J. R. and Lu, L.-H. (1991). Goodness-of-fit tests for regional generalized extreme value flood distributions, *Water Resources Res.* **27**, 1765–1776.

Christopeit, N. (1994). Estimating parameters of an extreme value distribution by the method of moments, *J. Statist. Planning Inference* **41**, 173–186.

Clough, D. J. and Kotz, S. (1965). Extreme value distributions with a special queueing model application, *CORS J.* **3**, 96–109.

Cockrum, M. B., Larson, R. K. and Taylor, R. W. (1990). Distribution modeling and simulation studies in product flammability testing, *ASA Procs. Business Econom. Statist. Section* 387–391.

Cohen, J. P. (1982a). The penultimate form of approximation to normal extremes, *Adv. App. Probab.* **14**, 324–339.

Cohen, J. P. (1982b). Convergence rates for the ultimate and pentulimate approximations in extreme value theory, *Adv. Appl. Probab.* **14**, 833–854.

Cohen, J. P. (1986). Large sample theory for fitting an approximating Gumbel model to maxima, *Sankhyā* A, **48**, 372–392.

Cohen, J. P. (1988). Fitting extreme value distributions to maxima, *Sankhyā* A, **50**, 74–97.

Coles, S. G. (1991). Statistical methodology for the multivariate analysis of environmental extremes, Ph.D. Thesis, University of Sheffield.

Coles, S. G. (1993). Regional modelling of extreme storms via max–stable processes, *J. Roy. Statist. Soc.* B, **55**, 797–816.

Coles, S. G. and Pan, F. (1996). The analysis of extreme pollution levels: a case study, *J. Appl. Statist.* **23**, 333–348.

Coles, S. G. and Powell, E. A. (1995). Bayesian methods in extreme value modelling, *Procs. of the 50th Session of the I.S.I.*, paper 15.2, pp. 607–627.

Coles, S. G. and Powell, E. A. (1996). Bayesian methods in extreme value modelling: a review and new developments, *Internat. Statist. Rev.* **64**, 119–136.

Coles, S. G. and Tawn, J. A. (1990). Statistics of coastal flood prevention, *Philos. Trans. Roy. Soc. London* A, **332**, 457–476.

Coles, S. G. and Tawn, J. A. (1991). Modelling extreme multivariate events, *J. Roy. Statist. Soc.* B, **53**, 377–392.

Coles, S. G. and Tawn, J. A. (1994). Statistical methods for multivariate extremes: an application to structural design, *Appl. Statist.* **43**, 1–48.

Coles, S. G. and Tawn, J. A. (1995). Modelling extremes: a Bayesian approach.

Coles, S. G. and Tawn, J. A. (1996a). Modelling extremes of the areal rainfall process, *J. Roy. Statist. Soc.* B, **58**, 329–347.

Coles, S. G. and Tawn, J. A. (1996b). A Bayesian analysis of extreme rainfall data, *Appl. Statist.* **45**, 463–478.

Coles, S. G., Tawn, J. A. and Smith, R. L. (1994). A seasonal Markov model for extremely low temperatures, *Environmetrics* **5**, 221–239.

Coles, S. G. and Walshaw, D. (1994). Directional modelling of extreme wind speeds, *Appl. Statist.* **43**, 139–157.

Cottis, R. A., Laycock, P. J., Holt, D., Moir, S. A. and Scarf, P. A. (1987). The statistics of pitting of austenitic stainless steels in chloride solutions, in *Advances in Localized Corrosion*, 117–121, eds. H. Issacs and U. Bertocci, NACE.

Cramér, H. (1946). Mathematical Methods of Statistics, Princeton University Press.

Csörgő, S. (1984). Adaptive estimation of the parameters of stable laws. in *Colloquia Mathematical Society Janos Bolyai 36. Limit Theorems in Probability and Statistics*, 305–386, North-Holland.

Csörgő, S., Deheuvels, P. and Mason, D. M. (1985). Kernel estimates of the tail index of a distribution, *Ann. Statist.* **13**, 1050–1077.

Cunnane, C. and Nash, J. E. (1974). Bayesian estimation of frequency of hydrological events, in *Symposium on Mathematical Models in Hydrology*, Warsaw, 1971. IAHS-AISH Publication No. 100, UNESCO.

D'Agostino, R. B. and Stephens, M. A. (eds.) (1986). *Goodness-of-Fit Techniques*, Dekker.

Daniels, H. E. (1942). A property of the distribution of extremes, *Biometrika* **32**, 194–195.

Dasgupta, R. and Bhaumik, D. K. (1995). Upper and lower tolerance limits of atmospheric ozone level and extreme value distribution, *Sankhyā* B, **57**, 182–199.

David, H. A. (1981). *Order Statistics*, 2nd edn., John Wiley and Sons.

Davidovich, M. I. (1992). On convergence of the Weibull–Gnedenko distribution to the extreme value distribution, *Vestnik Akad. Nauk Belaruss, Ser. Mat.-Fiz.*, No. 1, Minsk, 103–106.

Davison, A. C. (1986). Approximate predictive likelihood, *Biometrika* **73**, 323–332.

Davison, A. C. and Smith, R. L. (1990). Models for exceedances over high thresholds (with discussion), *J. Roy. Statist. Soc.* B, **52**, 393–442.

De Haan, L. (1970). *On Regular Variation and Its Application to the Weak Convergence of Sample Extremes*, Mathematical Centre Tracts **32**, Mathematisch Centrum. Amsterdam.

De Haan, L. (1971). A form of regular variation and its application to the domain of attraction of the double exponential, *Z. Wahrsch. Geb.* **17**, 241–258.

De Haan, L. (1976). Sample extremes: an elementary introduction, *Statist. Neerlandica* **30**, 161–172.

De Haan, L. (1984). A spectral representation for max–stable processes, *Ann. Probab.* **12**, 1194–1204.

De Haan, L. (1985). Extremes in high dimensions: the model and some statistics, *Procs. of the 45th Session of the I.S.I.*, paper 26.3.

De Haan, L. (1990). Fighting the arch-enemy with mathematics, *Statist. Neerlandica* **44**, 45–68.

De Haan, L. (1994). Extreme value statistics, in *Extreme Value Theory and Applications*, 93–122, eds. J. Galambos *et al.*, Kluwer.

De Haan, L. and De Rondé, J. (1998). Sea and wind: multivariate extremes at work, *Extremes* **1**, 7–46.

De Haan, L. and Resnick, S. I. (1987). On regular variation of probability densities, *Stochastic Processes and Their Applications* **25**, 83–93.

De Haan, L. and Stadtmüller, U. (1992). Generalized regular variation of second order Technical Report 96281A, Erasmus University, Rotterdam.

Deheuvels, P. (1981). Univariate extreme values – theory and applications, *Proc. 43rd Session of the I.S.I.* **49**, *2*, Buenos Aires.

Deheuvels, P., Hausler, E. and Mason, D. M. (1988). Almost sure convergence of the Hill estimator, *Math. Procs. Cambridge Philos. Soc.* **104**, 371–381.

Deheuvels, P. and Tiago de Oliveira, J. (1989). On the non-parametric estimation of the bivariate extreme-value distributions, *Statist. Probab. Lett.* **8**, 315–323.

Dekkers, A. L. M. and de Haan, L. (1989). On the estimation of the extreme-value index and large quantile estimation, *Ann. Statist.* **17**, 1795–1832.

Dekkers, A. L. M., Einmahl, J. H. J. and de Haan, L. (1989). A moment estimator for the index of an extreme-value distribution, *Ann. Statist.* **17**, 1833–1855.

Diebold, F. X., Schuermann, T. and Stroughair, J. D. (1999). Pitfalls and opportunities in the use of extreme value theory in risk management. Draft Report.

Dietrich, D. and Hüsler, J. (1996). Minimum distance estimators in extreme value distributions, *Comm. Statist. Theory and Methods* **25**, 695–703.

Dixon, M. J. and Tawn, J. A. (1992). Trends in U.K. extreme sea-levels: a spatial approach, *Geophys. J. Internat.* **111**, 607–616.

Dixon, M. J. and Tawn, J. A. (1994). Extreme sea levels: modelling interactions between tide and surge, in *Statistics for the Environment II*, eds. V. Barnett and K. F. Turkman, 221–232, John Wiley and Sons.

Dixon, M. J. and Tawn, J. A. (1995a). A semiparametric model for multivariate extreme values. *Statist. Comput.* **5**, 215–225.

Dixon, M. J. and Tawn, J. A. (1995b). *Estimates of extreme sea conditions: extreme sea-levels at the UK A class sites; optimal site-by-site analyses and spatial analysis for the east coast.* Proudman Oceanographic Laboratory Internal Document No. 72.

Dixon, M. J. and Tawn, J. A. (1998). The impact of non-stationarity in extreme sea-level estimation, *Appl. Statist.* **48**, 135–151.

Dodd, E. L. (1923). The greatest and least variate under general laws of error, *Trans. Amer. Math. Soc.* **25**, 525–539.

Doganaksoy, N. and Schmee, J. (1991). Comparisons of approximate confidence intervals for the smallest extreme value distribution simple linear regression model under time censoring, *Comm. Statist. Comput. Simulation* **20**, 1085–1113.

Downton, F. (1966). Linear estimates of parameters in the extreme value distribution, *Technometrics* **8**, 3–17.

Drees, H. (1995). Refined Pickands estimators of the extreme value index, *Ann. Statist.* **23**, 2059–2080.

Dubey, S. D. (1966). Characterization theorems for several distributions and their applications, *J. Industrial Math.* **16**, 1–22.

Dubey, S. D. (1969). A new derivation of the logistic distribution, *Naval Res. Logist. Quart.* **16**, 37–40.

Dupuis, D. J. (1996). Estimating the probability of obtaining nonfeasible parameter estimates of the generalized extreme-value distribution, *J. Statist. Comput. Simulation* **56**, 23–38.

Dupuis, D. J. and Field, C. A. (1998). Robust estimation of extremes, *Canad. J. Statist.* **26**, 199–215.

Einmahl, J. H. J., De Haan, L. and Huang, X. (1993). Estimating a multidimensional extreme-value distribution, *J. Multivariate Anal.* **47**, 35–47.

Einmahl, J. H. J., De Haan, L. and Sinha, A. K. (1997). Estimating the spectral measure of an extreme value distribution, *Stochastic Processes and Their Applications* **70**, 143–171.

Eldredge, G. G. (1957). Analysis of corrosion pitting by extreme value statistics and its application to oil well tubing caliper surveys, *Corrosion*, **13**, 51–76.

Embrechts, P., Klüppelberg, C. and Mikosch, T. (1997). *Modelling Extremal Events for Insurance and Finance*, Springer-Verlag.

Engelhardt, M. and Bain, L. J. (1973). Some complete and censored results for the Weibull or extreme-value distribution, *Technometrics* **15**, 541–549.

Engelund, S. and Rackwitz, R. (1992). On predictive distribution functions for the three asymptotic extreme value distributions, *Structural Safety* **11**, 255–258.

Epstein, B. (1948). Application to the theory of extreme values in fracture problems, *J. Amer. Statist. Assoc.* **43**, 403–412.

Epstein, B. (1960). Elements of the theory of extreme values, *Technometrics* **2**, 27–41.

Epstein, B. and Brooks, H. (1948). The theory of extreme values and its implications in the study of the dielectric strength of paper capacitors, *J. Appl. Phys.* **19**, 544–550.

Escalantesandoval, C. A. (1998). Multivariate extreme value distribution with mixed Gumbel marginals, *J. Amer. Water Resources Assoc.* **34**, 321–333.

Fahmi, K. J. and Abbasi, J. N. Al. (1991). Application of a mixture distribution of extreme values to earthquake magnitudes in Iraq and conterminous regions, *Geophys. J. Roy. Astron. Soc.* **107**, 209–217.

Falk, M. and Marohn, F. (1993). Von Mises conditions revisited, *Ann. Probab.* **21**, 1310–1328.

Fei, H., Kong, F. and Tang, Y. (1994). Estimations for two-parameter Weibull distributions and extreme-value distributions under multiple type-II censoring, *preprint.*

Feller, W. (1971). *An Introduction to Probability Theory and Its Applications.* Vol. 2, 2nd edition, John Wiley and Sons.

Finkelstein, B. V. (1953). On the limiting distributions of extreme terms of a variational series of a two-dimensional quantity, *Dok. Akad. Nauk* (N.S.) **91**, 209–210.

Fisher, R. A. (1934). Two new properties of mathematical likelihood, *Procs. Roy. Soc. London* A, **144**, 285–307.

Fisher, R. A. and Tippett, L. H. C. (1928). Limiting forms of the frequency distribution of the largest or smallest member of a sample, *Procs. Cambridge Philos. Soc.* **24**, 180–190.

Flood Studies Report (1975), see Jenkinson, A. F. (ed.) (1975).

Fréchet, M. (1927). Sur la loi de probabilité de l'écart maximum, *Ann. Soc. Polon. Math. Cracovie* **6**, 93–116.

Frenkel, J. I. and Kontorova, T. A. (1943). A statistical theory of the brittle strength of real crystals, *J. Phys. USSR* **7**, 108–114.

Fuller, W. E. (1914). Flood flows, *Trans. Amer. Soc. Civil Engineers* **77**, 564.

Fung, K. Y. and Paul, S. R. (1985). Comparisons of outlier detection procedures in Weibull or extreme-value distribution, *Comm. Statist. Simulation Comput.* **14**, 895–917.

Galambos, J. (1981a). Extreme value theory in applied probability, *Math. Scient.*

Galambos, J. (1981b). Failure time distributions: estimates and asymptotic results, in *Statistical Distributions in Scientific Work*, **5**, 309–317, eds. C. Taillie, G. P. Patil and B. A. Baldessari, Reidel.

Galambos, J. (1982). A statistical test for extreme value distributions, in *Nonparametric Statistical Inference*, 221–230, eds. B. V. Gnedenko, M. L. Puri and I. Vincze, North-Holland.

Galambos, J. (1987). *The Asymptotic Theory of Extreme Order Statistics*, 2nd edition, Krieger (1st edition, John Wiley and Sons, 1978).

Galambos, J. *et al.* (1994). Proceedings of the Gaithersburg Conference, 1993. Vol. 1: Kluwer; Vol. 2: Journal Research NIST; Vol. 3: NIST Special Publication 866.

Geffroy, J. (1958). Contribution à la théorie des valeurs extrêmes, *Publ. l'Inst. Statist. l'Univ. Paris* **7**, 37–121.

Geffroy, J. (1959). Contribution à la théorie des valeurs extrêmes, II, *Publ. l'Inst. Statist. l'Univ. Paris* **8**, 3–65.

Gerisch, W., Struck, W. and Wilke, B. (1991). One-sided Monte–Carlo tolerance limit factors for the exact extreme-value distributions from a normal parent distributional, *Comput. Statist. Quart.* **6**, 241–261.

Ghoudi, K., Khoudraji, A. and Rivest, L. P. (1998). Statistical properties of couples of bivariate extreme-value copulas, *Canad. J. Statist.* **26**, 187–197.

Gnedenko, B. (1943). Sur la distribution limite du terme maximum d'une série aléatoire, *Ann. Math.* **44**, 423–453. Translated and reprinted in: *Breakthroughs in Statistics*, Vol. **I**, 1992, eds. S. Kotz and N. L. Johnson, Springer-Verlag, pp. 195–225.

Goka, T. (1993). Application of extreme-value theory to reliability physics of electronic parts and to on-orbit single event phenomena. Paper presented at the *Conference on Extreme Value Theory and Its Applications*, May 2–7, 1993, National Institute of Standards, Gaithersburg.

Goldstein, N. (1963). Random numbers from the extreme value distribution, *Publ. l'Inst. Statist. l'Univ. Paris* **12**, 137–158.

Gomes, M. I. (1984). Penultimate limiting forms in extreme value theory, *Ann. Inst. Statist. Math.* **36**, 71–85.

Green, R. F. (1975). Consistent estimation based on extremes, *Technical Report No. 25*, University of California, Riverside.

Green, R. F. (1976). Partial attraction of maxima, *J. Appl. Probab.* **13**, 159–163.

Greenwood, J. A., Landwehr, J. M., Matalas, N. C. and Wallis, J. R. (1979). Probability weighted moments: Definition and relation to parameters of several distributions expressible in inverse form, *Water Resources Res.* **15**, 1049–1054.

Greenwood, M. (1946). The statistical study of infectious diseases, *J. Roy. Statist. Soc.* A, **109**, 85–109.

Greig, M. (1967). Extremes in a random assembly, *Biometrika* **54**, 273–282.

Greis, N. P. and Wood, E. F. (1981). Regional flood frequency estimation and network design, *Water Resources Res.* **17**, 1167–1177. (Correction, ibid. **19** (2), 589–590, 1983.)

Griffith, A. A. (1920). The phenomena of rupture and flow in solids, *Philos. Trans. Roy. Soc. London* A, **221**, 163–198.

Gumbel, E. J. (1935). Les valeurs extrêmes des distributions statistiques, *Ann. l'Inst. Henri Poincaré* **4**, 115–158.

Gumbel, E. J. (1937a). Les intervalles extrêmes entre les émissions radioactives, *J. Phys. Radium* **8**, 446–452.

Gumbel, E. J. (1937b). La durée extrême de la vie humaine, *Actualités Scientifique et Industrielles*, Hermann et Cie.

Gumbel, E. J. (1941). The return period of flood flows, *Ann. Math. Statist.* **12**, 163–190.

Gumbel, E. J. (1944). On the plotting of flood discharges, *Trans. Amer. Geophys. Union* **25**, 699–719.

Gumbel, E. J. (1945). Floods estimated by probability methods, *Engrg. News-Record* **134**, 97–101.

Gumbel, E. J. (1947). The distribution of the range, *Ann. Math. Statist.* **18**, 384–412.

Gumbel, E. J. (1949a). *The Statistical Forecast of Floods*, Bulletin No. **15**, 1–21, Ohio Water Resources Board.

Gumbel, E. J. (1949b). Probability tables for the range, *Biometrika* **36**, 142–148.

Gumbel, E. J. (1953). Introduction, in *Probability Tables for the Analysis of Extreme-Value Data*, National Bureau of Standards, Applied Mathematics Series, vol. 22.

Gumbel, E. J. (1954). *Statistical Theory of Extreme Values and Some Practical Applications*, National Bureau of Standards, Applied Mathematics Series, vol. 33.

Gumbel, E. J. (1958). *Statistics of Extremes*, Columbia University Press.

Gumbel, E. J. (1961). Sommes et différences de valeurs extrèmes indépendentes, *Comp. Rend. l'Acad. Sci. Paris* **253**, 2838–2839.

Gumbel, E. J. (1962a). Statistical estimation of the endurance limit — an application of extreme-value theory, in *Contributions to Order Statistics*, 406–431, eds. A. E. Sarhan and B. G. Greenberg, John Wiley and Sons.

Gumbel, E. J. (1962b). Statistical theory of extreme values (main results), in *Contributions to Order Statistics*, Chapter 6, eds. A. E. Sarhan and B. G. Greenberg, John Wiley and Sons.

Gumbel, E. J. (1962c). Produits et quotients de deux plus grandes valeurs indépendantes, *Comp. Rend. l'Acad. Sci. Paris*, **254**, 2132–2134.

Gumbel, E. J. (1962d). Produits et quotients de deux plus petites valeurs indépendantes, *Publ. l'Inst. Statist. l'Univ. Paris* **11**, 191–193.

Gumbel, E. J. (1962e). Multivariate extremal distributions, *Procs. Session of the I.S.I.* **39**, 471–475.

Gumbel, E. J. and Goldstein, N. (1964). Empirical bivariate extremal distributions, *J. Amer. Statist. Assoc.* **59**, 794–816.

Gumbel, E. J. and Herbach, L. H. (1951). The exact distribution of extremal quotient, *Ann. Math. Statist.* **22**, 418.

Gumbel, E. J. and Keeney, R. D. (1950). The extremal quotient, *Ann. Math. Statist.* **21**, 523.

Gumbel, E. J. and Mustafi, C. K. (1966). Comments to "The application of extreme value theory to error free communication", by Edward C. Posner, *Technometrics* **8**, 363–366.

Gumbel, E. J. and Mustafi, C. K. (1967). Some analytical properties of bivariate extremal distributions, *J. Amer. Statist. Assoc.* **62**, 569–588.

Gumbel, E. J. and Pickands, J. (1967). Probability tables for the extremal quotient, *Ann. Math. Statist.* **38**, 1541–1551.

Haeusler, E. and Teugels, J. L. (1985). On asymptotic normality of Hill's estimator for the exponent of regular variation, *Ann. Statist.* **13**, 743–756.

Hald, A. (1952). *Statistical Theory With Engineering Applications*, John Wiley and Sons.

Hall, M. J. (1992). Problems of handling messy field data for engineering decision-making: More on flood frequency analysis, *Math. Scientist* **17**, 78–88.

Hall, P. (1979). On the rate of convergence of normal extremes, *J. Appl. Probab.* **16**, 433–439.

Hall, P. (1980). Estimating probabilities for normal extremes, *Adv. Appl. Probab.* **12**, 491–500.

Harris, B. (1970). An application of extreme value theory to reliability theory, *Ann. Math. Statist.* **41**, 1456–1465.

Harter, H. L. (1970). *Order Statistics and Their Use in Testing and Estimation*, Vol. 2, Washington.

Harter, H. L. (1978). A bibliography of extreme-value theory, *Internat. Statist. Rev.* **46**, 279–306.

Harter, H. L. and Moore, A. H. (1968a). Maximum likelihood estimation, from doubly censored samples, of the parameters of the first asymptotic distribution of extreme values, *J. Amer. Statist. Assoc.* **63**, 889–901.

Hasofer, A. M. and Wang, Z. (1992). A test for extreme value domain of attraction, *J. Amer. Statist. Assoc.* **87**, 171–177.

Hassanein, K. M. (1965). *Estimation of the parameters of the extreme value distribution by order statistics.* National Bureau of Standards, Project No. 2776-M.

Hassanein, K. M. (1968). Analysis of extreme-value data by sample quantiles for very large samples, *J. Amer. Statist. Assoc.* **63**, 877–888.

Hassanein, K. M. (1969). Estimation of the parameters of the extreme value distribution by use of two or three order statistics, *Biometrika* **56**, 429–436.

Hassanein, K. M. (1972). Simultaneous estimation of the parameters of the extreme value distribution by sample quantiles, *Technometrics* **14**, 63–70.

Hassanein, K. M. and Saleh, A. K. Md. E. (1992). Testing equality of locations and quantiles of several extreme-value distributions by use of few order statistics of samples from extreme-value and Weibull distributions, in *Order Statistics and Nonparametrics: Theory and Applications*, 115–132, eds. P. K. Sen and I. A. Salama, North-Holland.

Henery, R. J. (1984). An extreme-value model for predicting the results of horse races, *Appl. Statist.* **33**, 125–133.

Hill, B. M. (1975). A simple general approach to inference about the tail of a distribution, *Ann. Statist.* **3**, 1163–1174.

Hisel, K. W. (ed.) (1994). *Extreme Values: Floods and Droughts*, Proc. Internat. Conf. on Stochastic and Statistical Methods in Hydrology and Environmental Engineering, Vol. 1, 1993, Kluwer.

Hooghiemstra, G. and Husler, J. (1996). A note on maxima of bivariate random vectors, *Statist. Probab. Lett.* **31**, 1–6.

Hopke, P. K. and Paatero, P. (1993). Extreme value estimation applied to aerosol size distributions and related environmental problems, paper presented at the *Conference on Extreme Value Theory and Its Applications*, May 2–7, 1993, National Institute of Standards.

Hosking, J. R. M. (1984). Testing whether the shape parameter is zero in the generalized extreme-value distribution, *Biometrika* **71**, 367–374.

Hosking, J. R. M. (1985). Maximum-likelihood estimation of the parameters of the generalized extreme-value distribution, *Appl. Statist.* **34**, 301–310.

Hosking, J. R. M. (1986). The theory of probability weighted moments, *Res. Rep. PC12210*, IBM Research. Reissued with corrections, 3 April 1989.

Hosking, J. R. M. and Wallis, J. R. (1987). Parameter and quantile estimation for the generalized Pareto distribution, *Technometrics* **29**, 339–349.

Hosking, J. R. M. and Wallis, J. R. (1988). The effect of intersite dependence on regional flood frequency analysis, *Water Resources Res.* **24**, 588–600.

Hosking, J. R. M., Wallis, J. R. and Wood, E. F. (1985a). An appraisal of the regional flood frequency procedure in the UK Flood Studies Report, *Hydrol. Sci. J.* **30**, 85–109.

Hosking, J. R. M., Wallis, J. R. and Wood, E. F. (1985b). Estimation of the generalized extreme-value distribution by the method of probability-weighted moments, *Technometrics* **27**, 251–261.

Hougaard, P. (1986). A class of multivariate failure time distributions, *Biometrika* **73**, 671–678.

Hüsler, J. and Schuepbach, M. (1986). On simple block estimators for the parameters of the extreme-value distribution, *Comm. Statist. Simulation Comput.* **15**, 61–76.

Hüsler, J. and Reiss, R. D. (1989). Maxima of normal random vectors: between independence and complete dependence, *Statist. Probab. Lett.* **7**, 283–286.

Jain, D. and Singh, V. P. (1987). Estimating parameters of EV1 distribution for flood frequency analysis, *Water Resources Res.* **23**, 59–71.

Jenkinson, A. F. (1955). Frequency distribution of the annual maximum (or minimum) values of meteorological elements, *Quart. J. Roy. Meteor. Soc.* **81**, 158–171.

Jenkinson, A. F. (1969). Statistics of Extremes, *Technical Note No. 98*, World Meteorological Organization. Chapter 5, 183–227.

Jenkinson, A. F. (Ed.) (1975). Natural Environment Research Council, *Flood Studies Rep.* **1**. London: NERC.

Jeruchim, M. C. (1976). On the estimation of error probability using generalized extreme value theory, *IEEE Trans. Information Theory* **IT-22**, 108–110.

Joe, H. (1990). Families of min–stable multivariate exponential and multivariate extreme value distributions *Statist. Probab. Lett.* **9**, 75–81.

Joe, H. (1994). Multivariate extreme value distributions with applications to environmental data, *Canad. J. Statist.* **22**, 47–64.

Joe, H. and Hu, T. H. (1996). Multivariate distributions from mixtures of max-infinitely divisible distributions, *J. Multivariate Anal.* **57**, 240–265.

Joe, H., Smith, R. L. and Weissmann, I. (1992). Bivariate threshold methods for extremes, *J. Roy. Statist. Soc.* B, **54**, 171–183.

Johns, M. V., Jr., and Lieberman, G. J. (1966). An exact asymptotically efficient confidence bound for reliability in the case of the Weibull distribution, *Technometrics* **8**, 135–175.

Johnson, N. L. and Kotz, S. (1972). *Distributions in Statistics: Continuous Multivariate Distributions*, John Wiley and Sons.

Johnson, N. L., Kotz, S. and Balakrishnan, N. (1995). *Continuous Univariate Distributions*, Vol. 2, 2nd edition, John Wiley and Sons.

Juncosa, M. L. (1949). The asymptotic behavior of the minimum in a sequence of random variables, *Duke Math. J.* **16**, 609–618.

Kanda, J. (1993). Application of an empirical extreme value distribution to load models, paper presented at the *Conference on Extreme Value Theory and Its Applications*, May 2–7, 1993, National Institute of Standards.

Kase, S. (1953). A theoretical analysis of the distribution of tensile strength of vulcanized rubber, *J. Polymer Sci.* **11**, 425–431.

Keating, J. P. (1984). A note on estimation of percentiles and reliability in the extreme-value distribution, *Statist. Probab. Lett.* **2**, 143–146.

Kimball, B. F. (1955). Practical applications of the theory of extreme values, *J. Amer. Statist. Assoc.* **50**, 517–528. (Correction: **50**, 1332.)

Kimball, B. F. (1956). The bias in certain estimates of the extreme-value distribution, *Ann. Math. Statist.* **27**, 758–767.

Kimball, B. F. (1960). On the choice of plotting positions on probability paper, *J. Amer. Statist. Assoc.* **55**, 546–560.

Kimber, A. C. (1985). Tests for the exponential, Weibull and Gumbel distributions based on the standardized probability plot, *Biometrika* **72**, 661–663.

King, J. R. (1959). Summary of extreme-value theory and its relation to reliability analysis, *Proc. 12th Annual Conference of the American Society for Quality Control*, **13**, 163–167.

Kinnison, R. (1989). Correlation coefficient goodness-of-fit test for the extreme-value distribution, *Amer. Statist.* **43**, 98–100.

Kinnison, R. P. (1985). *Applied Extreme Value Statistics*, Battelle Press, Macmillan.

Kluppelberg, C. and May, A. (1999). The dependence function for bivariate extreme value distributions – a systematic approach. Submitted for publication.

Landwehr, J. M., Matalas, N. C. and Wallis, J. R. (1979). Probability weighted moments compared with some traditional techniques in estimating Gumbel parameters and quantiles, *Water Resources Res.* **15**, 1055–1064.

Lawless, J. F. (1973). On the estimation of safe life when the underlying life distribution is Weibull, *Technometrics* **15**, 857–865.

Lawless, J. F. (1975). Construction of tolerance bounds for the extreme-value and Weibull distributions, *Technometrics* **17**, 255–262.

Lawless, J. F. (1978). Confidence interval estimation for the Weibull and extreme value distributions (with discussion), *Technometrics* **20**, 355–368.

Lawless, J. F. (1980). Inference in the generalized gamma and log-gamma distribution, *Technometrics* **22**, 67–82.

Lawless, J. F. (1982). *Statistical Models & Methods for Lifetime Data*, John Wiley and Sons.

Leadbetter, M. R., Lindgren, G. and Rootzén, H. (1983). *Extremes and Related Properties of Random Sequences and Processes*, Springer-Verlag.

Ledford, A. W. and Tawn, J. A. (1996). Statistics for near independence in multivariate extreme values, *Biometrika* **83**, 169–187.

Ledford, A. W. and Tawn, J. A. (1997). Modelling dependence within joint tail regions, *J. Roy. Statist. Soc.* B, **59**, 475–499.

Lettenmaier, D. P. and Burges, S. J. (1982). Gumbel's extreme value 1 distribution, a new look, *J. Hydraulics Division, Proc. ASCE* **108**, 502–514.

Lettenmaier, D. P., Wallis, J. R. and Wood, E. F. (1987). Effect of regional heterogeneity on flood frequency estimation, *Water Resources Res.* **23**, 313–323.

Levy, P. (1939). Sur la division d'un segment par des points choisis au hasard, *C. R. Acad. Sci. Paris* **208**, 147–149.

Lieblein, J. (1953). On the exact evaluation of the variances and covariances of order statistics in samples from the extreme-value distribution, *Ann. Math. Statist.* **24**, 282–287.

Lieblein, J. (1962). Extreme-value distribution, in *Contributions to Order Statistics*, 397–406, eds. A. E. Sarhan and B. G. Greenberg, John Wiley and Sons.

Lieblein, J. and Zelen, M. (1956). Statistical investigation of the fatigue life of deep-grove ball bearings, *J. Res. National Bureau of Standards* **57**, 273–316.

Lingappaiah, G. S. (1984). Bayesian prediction regions for the extreme order statistics, *Biom. J.* **26**, 49–56.

Lockhart, R. A., O'Reilly, F. J. and Stephens, M. A. (1986). Tests for the extreme value and Weibull distributions based on normalized spacings, *Naval Res. Logist. Quart.* **33**, 413–421.

Lockhart, R. A. and Spinelli, J. J. (1990). Comment on "Correlation coefficient goodness-of-fit tests for the extreme-value distribution". *Amer. Statist.* **44**, 259–260.

Longuet-Higgins, M. S. (1952). On the statistical distribution of the heights of sea waves, *J. Marine Res.* **9**, 245–266.

Lowery, M. D. and Nash, J. E. (1970). A comparison of methods of fitting the double exponential distribution, *J. Hydrol.* **10**, 259–275.

Lu, L.-H. and Stedinger, J. R. (1992). Variance of two- and three-parameter GEV/PWM quantile estimators: Formulae, confidence intervals, and a comparison, *J. Hydrol.* **138**, 247–267.

Luceño, A. (1994). Speed of convergence to the extreme value distributions on their probability plotting papers, *Comm. Statist. B*, **23**, 529–545.

Macleod, A. J. (1989). Comment on "Maximum-likelihood estimation of the parameters of the generalized extreme-value distribution", *Appl. Statist.* **38**, 198–199.

Mahmoud, M. W. and Ragab, A. (1975). On order statistics in samples drawn from the extreme value distributions, *Math. Operationsforschung Statist Series Statist.* **6**, 809–816.

Mann, N. R. (1967a). *Results on Location and Scale Parameter Estimation with Application to the Extreme-Value Distribution*, Report ARL67-0023, Aerospace Research Laboratories.

Mann, N. R. (1967b). Tables for obtaining the best linear invariant estimates of the parameters of the Weibull distribution, *Technometrics* **9**, 629–645.

Mann, N. R. (1969). Optimum estimators for linear functions of location and scale parameters, *Ann. Math. Statist.* **40**, 2149–2155.

Mann, N. R. and Fertig, K. W. (1973). Tables for obtaining Weibull confidence bounds and tolerance bounds based on best linear invariant estimates of parameters of the extreme-value distribution, *Technometrics* **15**, 87–102.

Mann, N. R. and Fertig, K. W. (1975). Simplified efficient point and interval estimators for Weibull parameters, *Technometrics* **17**, 361–368.

Mann, N. R. and Fertig, K. W. (1977). Efficient unbiased quantile estimators for moderate-size complete samples from extreme-value and Weibull distributions: Confidence bounds and tolerance and prediction intervals, *Technometrics* **19**, 87–94.

Mann, N. R., Schafer, R. E. and Singpurwalla, N. D. (1974). *Methods for Statistical Analysis of Reliability and Life Data*, John Wiley and Sons.

Mann, N. R., Scheuer, E. M. and Fertig, K. W. (1973). A new goodness-of-fit test for the two parameter Weibull or extreme-value distribution with unknown parameters, *Comm. Statist.* **2**, 383–400.

Mann, N. R. and Singpurwalla, N. D. (1982). Extreme-value distributions, in *Encyclopedia of Statistical Sciences*, **2**, 606–613, eds. S. Kotz, N. L. Johnson and C. B. Read, John Wiley and Sons.

Marco, J. M. and Ruiz-Rivas, C. (1992). On the construction of multivariate distributions with given nonoverlapping multivariate marginals, *Statist. Probab. Lett.* **15**, 259–265.

Marcus, M. B. and Pinsky, M. (1969). On the domain of attraction of $\exp(-e^{-x})$, *J. Math. Anal. Appl.* **28**, 440–449.

Maritz, J. S. and Munro, A. H. (1967). On the use of generalized extreme-value distribution in estimating extreme percentiles, *Biometrics* **23**, 79–103.

Marshall, A. W. and Olkin, I. (1967). A multivariate exponential distribution, *J. Amer. Statist. Assoc.* **62**, 30–44.

Marshall, A. W. and Olkin, I. (1983). Domains of attraction of multivariate extreme value distributions. *Ann. Probab.* **11**, 168–177.

Marshall, R. J. (1983). A spatial-temporal model of storm rainfall, *J. Hydrol.* **62**, 53–62.

Mason, D. M. (1982). Laws of large numbers for sums of extreme values, *Ann. Probab.* **10**, 754–764.

McFadden, D. (1978). Modelling the choice of residential location, in *Spatial Interaction Theory and Planning Models*, eds. A. Karlqvist, L. Lundquist, F. Snickers and J. Weibull, 75–96, North-Holland.

McLaren, C. G. and Lockhart, R. A. (1987). On the asymptotic efficiency of certain correlation tests of fit, *Canad. J. Statist.* **15**, 159–167.

Mejzler, D. G. (1949). On a theorem of B. V. Gnedenko. *Sb. Trudov Inst. Mat. Akad. Nauk Ukrain. SSR* **12**, 31–35 (in Russian).

Mejzler, D. G. and Weissman, I. (1969). On some results of N. V. Smirnov concerning limit distributions for variational series, *Ann. Math. Statist.* **40(2)**, 480–491.

Metcalfe, A. G. and Smith, G. K. (1964). Effects of length on the strength of glass fibres, *Procs. Amer. Soc. Testing Materials* **64**, 1075–1093.

Metcalfe, A. V. and Mawdsley, J. A. (1981). Estimation of extreme low flows for pumped storage reservoir design, *Water Resources Res.* **17**, 1715–1721.

Michael, J. R. (1983). The stabilized probability plot, *Biometrika* **70**, 11–17.

Mises, R., von (1923). Über die Variationsbreite einer Beobachtungsreihe, *Sitzungsber. Berlin. Math. Ges.* **22**, 3–8.

Mises, R., von (1936). La distribution de la plus grande de n valeurs, *Rev. Math. Union Interbalk.* **1**, 141–160. Reproduced in *Selected Papers of Richard von Mises, II* (1954), pp. 271–294, Amer. Math. Soc.

Moran, P. A. P. (1947). The random division of an interval, *J. Roy. Statist. Soc.* B, **40**, 213–216.

Moran, P. A. P. (1953). The random division of an interval — Part III, *J. Roy. Statist. Soc.* B, **15**, 77–80.

Morton, I. D. and Bowers, J. (1996). Extreme value analysis in a multivariate offshore environment, *Appl. Ocean Res.* **18**, 303–317.

Mustafi, C. K. (1963). Estimation of parameters of the extreme value distribution with limited type of primary probability distribution, *Bull. Calcutta Statist. Assoc.* **12**, 47–54.

Nadarajah, S. (1994). Multivariate extreme value methods with applications to reservoir flood safety, Ph.D. Thesis, University of Sheffield.

Nadarajah, S. (1999a). A polynomial model for bivariate extreme value distributions, *Statist. Probab. Lett.* **42**, 15–25.

Nadarajah, S. (1999b). Simulation of multivariate extreme values, *J. Statist. Comput. Simulation* **62**, 395–410.

Nadarajah, S. (1999c). Multivariate extreme value distributions based on bivariate structures, Unpublished Technical Note.

Nadarajah, S. (2000). Extreme value models: univariate and multivariate, to appear in *Handbook of Statistics*, eds. C. R. Rao and D. N. Shanbhag, John Wiley and Sons.

Nadarajah, S., Anderson, C. W. and Tawn, J. A. (1998). Ordered multivariate extremes, *J. Roy. Statist. Soc.* B, **60**, 473–496.

Nagaraja, H. N. (1982). Record values and extreme value distributions, *J. Appl. Probab.* **19**, 233–239.

Nagaraja, H. N. (1984). Asymptotic linear prediction of extreme order statistics, *Ann. Inst. Statist. Math.* **36**, 289–299.

Nissan, E. (1988). Extreme value distribution in estimation of insurance premiums, *ASA Procs. Business Econom. Statist. Sec.* 562–566.

Niu, X. F. (1997). Extreme value theory for a class of nonstationary time series with applications, *Ann. Appl. Probab.* **7**, 508–522.

Nordquist, J. M. (1945). Theory of largest values, applied to earthquake magnitudes, *Trans. Amer. Geophys. Union* **26**, 29–31.

Oakes, D. and Manatunga, A. K. (1992). Fisher information for a bivariate extreme value distribution, *Biometrika* **79**, 827–832.

Obretenov, A. (1991). On the dependence function of Sibuya in multivariate extreme value theory, *J. Multivariate Anal.* **36**, 35–43.

Ogawa, J. (1951). Contributions to the theory of systematic statistics, I, *Osaka Math. J.* **3**, 175–213.

Ogawa, J. (1952). Contributions to the theory of systematic statistics, II, *Osaka Math. J.* **4**, 41–61.

Okubo, T. and Narita, N. (1980). On the distribution of extreme winds expected in Japan, *National Bureau of Standards Special Publ.* **560–1**, 12 pp.

Otten, A. and Montfort, M. A. J., van (1980). Maximum likelihood estimation of the general extreme-value distribution parameters, *J. Hydrol.* **47**, 187–192.

Owen, D. B. (1962). *Handbook of Statistical Tables*, Addison-Wesley.

Öztürk, A. (1986). On the W test for the extreme value distribution, *Biometrika* **73**, 738–740.

Öztürk, A. and Korukoğlu, S. (1988). A new test for the extreme value distribution, *Comm. Statist. Simulation Comput.* **17**, 1375–1393.

Peirce, F. T. (1926). Tensile tests for cotton yarns v. 'the weakest link' — Theorems on the strength of long and of composite specimens, *J. Textile Inst. Tran.* **17**, 355.

Peng, L. (1999). Estimation of the coefficients of tail dependence in bivariate extremes, *Statist. Probab. Lett.* **43**, 399–409.

Pericchi, L. P. and Rodriguez-Iturbe, I. (1985). On the statistical analysis of floods, in *A Celebration of Statistics: The I.S.I. Centenary Volume*, 511–541, eds. A. C. Atkinson and S. E. Fienberg, Springer-Verlag.

Phien, H. N. (1991). Maximum likelihood estimation for the Gumbel distribution from censored samples, in *The Frontiers of Statistical Computation, Simulation, and Modeling*, **1**, 271–287, eds. P. R. Nelson, E. J. Dudewicz, A. Öztürk and E. C. van der Meulen, American Sciences Press.

Pickands, J. (1968). Moment convergence of sample extremes, *Ann. Math. Statist.* **39**, 881–889.

Pickands, J. (1971). The two-dimensional Poisson process and extremal processes, *J. Appl. Probab.* **8**, 745–756.

Pickands, J. (1975). Statistical inference using extreme order statistics, *Ann. Statist.* **3**, 119–131.

Pickands, J. (1981). Multivariate extreme value distributions, *Proc. 43rd Session of the ISI*, Buenos Aires, **49**, 859–878.

Posner, E. C. (1965). The application of extreme-value theory to error-free communication, *Technometrics* **7**, 517–529.

Potter, W. D. (1949). Normalcy tests of precipitation and frequency studies of runoff on small watersheds, *U.S. Department of Agriculture Technical Bulletin, No. 985.*

Prentice, R. L. (1974). A log gamma model and its maximum likelihood estimation, *Biometrika* **61**, 539–544.

Prescott, P. and Walden, A. T. (1980). Maximum likelihood estimation of the parameters of the generalized extreme-value distribution, *Biometrika* **67**, 723–724.

Prescott, P. and Walden, A. T. (1983). Maximum likelihood estimation of the parameters of the three-parameter generalized extreme-value distribution from censored samples, *J. Statist. Comput. Simulation* **16**, 241–250.

Press, H. (1949). The application of the statistical theory of extreme value to gust-load problems, *National Advisory Committee on Aeronautics, Technical Note No. 1926.*

Provasi, C. (1987). Exact and approximate means and covariances of order statistics of the standardized extreme value distribution (I type), *Riv. Statist. Appl.* **20**, 287–295. (In Italian.)

Pugh, D. T. (1982). Estimating extreme currents by combining tidal and surge probabilities, *Ocean Engrg.* **9**, 361–372.

Pugh, D. T. and Vassie, J. M. (1980). Applications of the joint probability method for extreme sea-level computations, *Procs. Inst. Civil Engineers* **69**, 959–975.

Pyke, R. (1965). Spacings (with discussion), *J. Roy. Statist. Soc.* B, **27**, 395–449.

Pyle, J. A. (1985). Assessment models, in *Atmospheric Ozone*, Chap. 12. NASA.

Qi, Y. G. (1998). Estimating extreme-value index from records, *Chin. Ann. Math.* **B19**, 4, 499–510.

Rajan, K. (1993). Extreme value theory and its applications in microstructural sciences, paper presented at the *Conference on Extreme Value Theory and Its Applications*, May 2–7, 1993, National Institute of Standards.

Rantz, S. F. and Riggs, H. C. (1949). *Magnitude and frequency of floods in the Columbia river basin*, U.S. Geological Survey, Water Supply Paper 1080, 317–476.

Rasheed, H., Aldabagh, A. S. and Ramamoorthy, M. V. (1983). Rainfall analysis by power transformation, *J. Climate Appl. Meteor.* **22**, 1411–1415.

Reiss, R.-D. (1989). *Approximate Distributions of Order Statistics: With Applications to Nonparametric Statistics*, Springer-Verlag.

Reiss, R.-D. and Thomas, M. (1997). *Statistical Analysis of Extreme Values*, Birkhäuser.

Resnick, S. I. (1987). *Extreme Values, Regular Variation and Point Processes*, Springer-Verlag.

Revfeim, K. J. A. (1984a). The cumulants of an extended family of type I extreme value distributions, *Sankhyā* B, **46**, 281–284.

Revfeim, K. J. A. (1984b). Generating mechanisms of, and parameter estimators for, the extreme value distribution, *Austral. J. Statist.* **26**, 151–159.

Revfeim, K. J. A. and Hessell, J. W. D. (1984). More realistic distributions for extreme wind gusts, *Quart. J. Roy. Meteor. Soc.* **110**, 505–514.

Ripley, B. D. (1987). *Stochastic Simulation*, John Wiley and Sons.

Robinson, M. E. and Tawn, J. A. (1995). Statistics for exceptional athletic records, *Appl. Statist.* **44**, 499–511.

Robinson, M. E. and Tawn, J. A. (1997). Statistics for extreme sea currents, *Appl. Statist.* **46**, 183–205.

Roldan-Canas, J., Garcia-Guzman, A. and Losada-Villasante, A. (1982). A stochastic model for wind occurrence, *J. Appl. Meteor.* **21**, 740–744.

Rootzén, H. (1978). Extremes of moving averages of stable processes, *Ann. Probab.* **6**, 847–869.

Rosengard, A. (1962). Etude des lois limites jointes et marginales de la mayonne et des valeurs extrêmes, *Publ. ISUP* **XI**, **1**, 1–56.

Rossi, F. (1986). Reply to "Comment on 'Two-component extreme value distribution for flood frequency analysis'", *Water Resources Res.* **22**, 267–269.

Rossi, F., Fiorentino, M. and Versace, P. (1986). Two-component extreme value distribution for flood frequency analysis, *Water Resources Res.* **22**.

Scarf, P. A. (1992). Estimation for a four parameter generalized extreme value distribution, *Comm. Statist. Theory and Methods* **21**, 2185–2201.

Scarf, P. A. and Laycock, P. J. (1993). Applications of extreme value theory in corrosion engineering. paper presented at the *Conference on Extreme Value Theory and Its Applications*, May 2–7, 1993, National Institute of Standards.

Scarf, P. A. and Laycock, P. J. (1996). Estimation of extremes in corrosion engineering, *J. Appl. Statist.* **23**, 621–643.

Schuepbach, M. and Huesler, J. (1983). Simple estimators for the parameters of the extreme-value distribution based on censored data, *Technometrics* **25**, 189–192.

Schrupp, K. and Rackwitz, R. (1984). Conjugate priors in extreme value theory, in: Prädiktive Verteilungen and ihre Anwendungen in der Zuverlässigheitstheorie der Bauwerke, *Berichte zur Zuverlässigheitstheorie der Bauwerke*, LKI, Technical University Munich. Heft **71**.

Sen, P. K. (1961). A note on the large-sample behaviour of extreme sample values from distributions with finite end-points, *Bull. Calcutta Statist. Assoc.* **10**, 106–115.

Sethuraman, J. (1965). On a characterization of the three limiting types of the extreme, *Sankhyā*, A, **27**, 357–364.

Shapiro, S. S. and Brain, C. W. (1987). W-test for the Weibull distribution, *Comm. Statist.* **B16**, 209–219.

Shen, H. W., Bryson, M. C. and Ochoa, I. D. (1980). Effect of tail behaviour assumptions on flood predictions, *Water Resources Res.* **16**, 361–364.

Shi, D. (1995a). Moment estimation for multivariate extreme value distributions, *J. Appl. Math.* **10B**, 61–68.

Shi, D. (1995b). Fisher information for a multivariate extreme value distribution, *Biometrika* **82**, 644–649.

Shi, D., Smith, R. L. and Coles, S. G. (1993). Joint versus marginal estimation for bivariate extremes, Unpublished Technical Report.

Shi, D. and Zhou, S. (1999). Moment estimation for multivariate extreme value distribution in a nested logistic model, *Ann. Inst. Statist. Math.* **51**, 253–264.

Shibata, T. (1993). Application of extreme value statistics to corrosion, paper presented at the *Conference on Extreme Value Theory and Its Applications*, May 2–7, 1993, National Institute of Standards.

Shimokawa, T. and Liao, M. (1999). Goodness-of-fit tests for type-I extreme value and 2-parameter Weibull distributions, *IEEE Trans. Reliab.* **48**, 79–84.

Sibuya, M. (1967). On exponential and other random variable generators, *Ann. Inst. Statist. Math.* **13**, 231–237.

Silverman, B. W. (1986). *Density Estimation for Statistics and Data Analysis*, Chapman and Hall.

Simiu, E., Bietry, J. and Filliben, J. J. (1978). Sampling errors in estimation of extreme winds, *J. Structural Div. National Bureau of Standards* **104**, 491–501.

Simiu, E. and Filliben, J. J. (1975). Statistical analysis of extreme winds, *National Bureau of Standards Technical Note* **868**, 52 pp.

Simiu, E. and Filliben, J. J. (1976). Probability distributions of extreme wind speeds, *J. Structural Div. National Bureau of Standards* **102**, 1861–1877.

Singpurwalla, N. D. (1972). Extreme values from a lognormal law with applications to air pollution problems, *Technometrics* **14**, 703–711.

Sivapalan, M. and Bloschl, G. (1998). Transformation of point rainfall to area rainfall: Intensity-duration frequency curves, *J. Hydrol.* **204**, 150–167.

Smirnov, N. V. (1949). Limit distributions for the terms of a variational series, *Trudy Mat. Inst. Stekl.* **25**, 1–60.

Smirnov, N. V. (1952). Limit distributions for the terms of a variational series, *Trans. Amer. Math. Soc.* Sec. 1, No. **67**, 1–64 (English translation).

Smith, J. A. (1987).[a] Estimating the upper tail of flood frequency distributions, *Water Resources Res.* **23**, 1657–1666.

Smith, R. L. (1985). Maximum likelihood estimation in a class of non-regular cases, *Biometrika* **72**, 67–90.

Smith, R. L. (1986). Extreme value theory based on the r largest annual events, *J. Hydrol.* **86**, 27–43.

Smith, R. L. (1987). Estimating tails of probability distributions, *Ann. Statist.* **15**, 1174–1207.

Smith, R. L. (1989). Extreme value analysis of environmental time series: an application to trend detection in ground-level ozone, *Statist. Sci.* **4**, 367–393.

Smith, R. L. (1991). Regional estimation from spatially dependent data.

Smith, R. L. (1992). Introduction to Gnedenko (1943). In *Breakthroughs in Statistics*, **I**, 185–194.

Smith, R. L. (1994). Multivariate threshold methods, in *Extreme Value Theory and Its Applications* 249–268, eds. J. Galambos, J. Lechner and E. Simiu, Kluwer.

Smith, R. L. and Naylor, J. C. (1987). A comparison of maximum likelihood and Bayesian estimators for the three-parameter Weibull distribution, *Appl. Statist.* **36**, 358–369.

Smith, R. L., Tawn, J. A. and Coles, S. G. (1997). Markov chain models for threshold exceedances, *Biometrika* **84**, 249–268.

[a]The reader should note that this bibliography contains 4 different Smiths.

Smith, R. L., Tawn, J. A. and Yuen, H. K. (1990). Statistics of multivariate extremes, *Internat. Statist. Rev.* **58**, 47–58.

Smith, R. L. and Weissman, I. (1985). Maximum likelihood estimation of the lower tail of a probability distribution, *J. Roy. Statist. Soc.* B, **47**, 285–298.

Smith, R. M. (1977). Some results on interval estimation for the two parameter Weibull or extreme-value distribution, *Comm. Statist. Theory and Methods* **2**, 1311–1322.

Smith, R. M. and Bain, L. J. (1976). Correlation type of goodness-of-fit statistics with censored sampling, *Comm. Statist. Theory and Methods* **5**, 119–132.

Smith, T. E. (1984). A choice of probability characterization of generalized extreme value models, *Appl. Math. Comput.* **14**, 35–62.

Stedinger, J. R., Vogel, R. M. and Foufoula-Geoorgiou, E. (1993). Frequency analysis of extreme events, *Handbook of Hydrology*, Chap. 18, ed. D. R. Maidment, McGraw-Hill.

Stephens, M. A. (1977). Goodness of fit for the extreme value distribution, *Biometrika* **64**, 583–588.

Stephens, M. A. (1986). Tests based on regression and correlation, in *Goodness-of-fit Techniques*, Chap. 5, eds. R. B. D'Agostino and M. A. Stephens, Dekker.

Stone, G. C. and Rosen, H. (1984). Some graphical techniques for estimating Weibull confidence intervals, *IEEE Trans. Reliab.* **33**, 362–369.

Sweeting, T. J. (1985). On domains of uniform local attraction in extreme value theory, *Ann. Probab.* **13**, 196–205.

Takahashi, R. (1987). Some properties of multivariate extreme value distributions and multivariate tail equivalence, *Ann. Inst. Statist. Math.* A, **39**, 637–647.

Takahashi, R. (1994a). Domains of attraction of multivariate extreme value distributions, *J. Res. National Inst. Standards Tech.* **99**, 551–554.

Takahashi, R. (1994b). Asymptotic independence and perfect dependence of vector components of multivariate extreme statistics, *Statist. Probab. Lett.* **19**, 19–26.

Tawn, J. A. (1988a). An extreme value theory model for dependent observations, *J. Hydrol.* **101**, 227–250.

Tawn, J. A. (1988b). Bivariate extreme value theory: Models and estimation, *Biometrika* **75**, 397–415.

Tawn, J. A. (1990). Modelling multivariate extreme value distributions, *Biometrika* **77**, 245–253.

Tawn, J. A. (1992). Estimating probabilities of extreme sea-levels, *Appl. Statist.* **41**, 77–93.

Tawn, J. A. (1994). Applications of multivariate extremes, in *Extreme Value Theory and Its Applications*, eds. J. Galambos, J. Lechner and E. Simiu, Kluwer.

Tawn, J. A. and Vassie, J. M. (1989). Extreme sea-levels: the joint probabilities method revisited and revised, *Procs. Inst. Civil Engineers*, part 2, **87**, 429–442.

Taylor, R. W. (1991). The development of burn time models to simulate product flammability testing, *ASA Procs. Business Econom. Statist. Sec.* 339–344.

Teugels, J. L. and Beirlant, J. (1993). Extremes in insurance, Paper presented at the *Conference on Extreme Value Theory and Its Applications*, May 2–7, 1993, NIST, Gaitershurg, MD.

Thom, H. C. S. (1954). Frequency of maximum wind speeds, *Procs. Amer. Soc. Civil Engineers* **80**, 104–114.

Thomas, D. R., Bain, L. J. and Antle, C. E. (1970). Reliability and tolerance limits in the Weibull distribution, *Technometrics* **12**, 363–371.

Thomas, D. R. and Wilson, W. M. (1972). Linear order statistic estimation for the two-parameter Weibull and extreme-value distributions from type II progressively censored samples, *Technometrics* **14**, 679–691.

Tiago de Oliveira, J. (1958). Extremal distributions, *Rev. Fac. Ciencias Lisboa, 2 ser., A, Mat., VII*, 215–227.

Tiago de Oliveira, J. (1962). The asymptotic independence of the sample mean and extremes, *Rev. Fac. Sci. Lisboa, A*, **VIII**, **2**, 299–310.

Tiago de Oliveira, J. (1963). Decision results for the parameters of the extreme value (Gumbel) distribution based on the mean and the standard deviation, *Trabajos Estadística* **14**, 61–81.

Tiago de Oliveira, J. (1972). Statistics for Gumbel and Fréchet distributions, in *Structural Safety and Reliab.*, 94–105, ed. A. Freudenthal, Pergamon.

Tiago de Oliveira, J. (1981). Statistical choice of univariate extreme models. In *Statistical Distributions in Scientific Work*, **6**, 367–387, eds. C. Taillie, G. P. Patil and B. A. Baldessari, Reidel.

Tiago de Oliveira, J. (1983). Gumbel distribution, in *Encyclopedia of Statistical Sciences*, **3**, 552–558, eds. S. Kotz, N. L. Johnson and C. B. Read, John Wiley and Sons.

Tiago de Oliveira, J. (1987). Intrinsic estimation of the dependence structure for bivariate extremes. *Technical Report*, 87–18, Dept of Statistics, Iowa State University.

Tiku, M. L. and Singh, M. (1981). Testing the two parameter Weibull distribution, *Comm. Statist. Theory and Methods* **10**, 907–918.

Tippett, L. H. C. (1925). On the extreme individuals and the range of samples taken from a normal population, *Biometrika* **17**, 364–387.

Tsujitani, M., Ohta, H. and Kase, S. (1979). A preliminary test of significance for the extreme-value distribution, *Bull. Univ. Osaka Prefecture* A, **27**, 187–193.

Tsujitani, M., Ohta, H. and Kase, S. (1980). Goodness-of-fit test for extreme-value distribution, *IEEE Trans. Reliab.* **29**, 151–153.

van Montfort, M. A. J. (1970). On testing that the distribution of extreme is of type I when Type-II is the alternative, *J. Hydrol.* **11**, 421–427.

van Montfort, M. A. J. (1973). An asymmetric test on the type of the distribution of extremes, *Med. Landbouwhogeschool* **73–18**, 1–15.

Vassie, J. M., Blackman, D. L., Tawn, J. A. and Dixon, M. J. (1999). *Extreme still water levels along the Humber Estuary*, Joint Report: Proudman Oceanographic Laboratory and Lancaster University.

Velz, C. J. (1947). Factors influencing self-purification and their relation to pollution abatement, *Sewage Works J.* **19**, 629–644.

Viveros, R. and Balakrishnan, N. (1994). Interval estimation of parameters of life from progressively censored data, *Technometrics* **36**, 84–91.

Vogel, R. M. (1986). The probability plot correlation coefficient test for the normal, lognormal, and Gumbel distributional hypotheses, *Water Resources Res.* **22**, 587–590.

Wallis, J. R. (1980). Risk and uncertainties in the evaluation of flood events for the design of hydraulic structures, in *Piene e Siccità*, 3–36, eds. E. Guggino, G. Rossi and E. Todini, Fondazione Politecnico del Mediter.

Wallis, J. R. and Wood, E. F. (1985). Relative accuracy of log Pearson III procedures, *J. Hydraul. Engrg.* **111**, 1043–1056.

Wang, J. Z. (1995). Selection of the k largest order statistics for the domain of attraction of the Gumbel distribution, *J. Amer. Statist. Assoc.* **90**, 1055–1061.

Wang, Q. J. (1990). Estimation of the GEV distribution from censored samples by method of partial probability weighted moments, *J. Hydrol.* **120**, 103–114.

Wantz, J. W. and Sinclair, R. E. (1981). Distribution of extreme winds in the Bonneville power administration service area, *J. Appl. Meteor.* **20**, 1400–1411.

Watabe, M. and Kitagawa, Y. (1980). Expectancy of maximum earthquake motions in Japan, *National Bureau of Standards Special Publ.* **560–10**, 8 pp.

Watson, G. S. (1954). Extreme values in samples from m-dependent stationary stochatsic processes, *Ann. Math. Statist.* **25**, 798–800.

Weibull, W. (1939a). A statistical theory of the strength of materials, *Ing. Vet. Akad. Handlingar* **151**.

Weibull, W. (1939b). The phenomenon of rupture in solids, *Ing. Vet. Akad. Handlingar* **153**, 2.

Weinstein, S. B. (1973). Theory and application of some classical and generalized asymptotic distributions of extreme values, *IEEE Trans. Information Theory* **IT-19**, 148–154.

Weiss, L. (1961). On the estimation of scale parameters, *Naval Res. Logist. Quart.* **8**, 245–256.

Weissman, I. (1978). Estimation of parameters and large quantiles based on the k largest observations, *J. Amer. Statist. Assoc.* **73**, 812–815.

Whittaker, J. (1990). *Graphical Models in Applied Multivariate Statistics*, John Wiley and Sons.

Wiggins, J. B. (1991). Empirical tests of the bias and efficiency of the extreme-value variance estimator for common stocks, *J. Business Univ. Chicago* **64**, 417–432.

Worms, R. (1998). Propriété asymptotique des excès additifs et valeurs extrêmes: le cas de la loi de Gumbel, *C. R. Acad. Sci. Paris,* t. 327, Série **I**, 509–514.

Xapsos, M. A., Summers, G. P. and Barke, E. A. (1998). Extreme value analysis of solar energetic motion peak fluxes, *Solar Phys.* **183**, 157–164.

Yasuda, T. and Mori, N. (1997). Occurrence properties of giant freak waves in sea area around Japan, *J. Waterway Port Coastal Ocean Engrg. — ASCE* **123**, 209–213.

Young, D. H. and Bakir, S. T. (1987). Bias correction for a generalized log-gamma regression model, *Technometrics* **29**, 183–191.

Yun, S. (1997). On domains of attraction of multivariate extreme value distributions under absolute continuity, *J. Multivariate Anal.* **63**, 277–295.

Zachary, S., Feld, G., Ward, G. and Wolfram, J. (1998). Multivariate extrapolation in the offshore environment, *Appl. Ocean Res.* **20**, 273–295.

Zelenhasic, E. (1970). Theoretical probability distributions for flood peaks, *Hydrol. Paper No. 42.* Fort Collins: Colorado State University.

Zelterman, D. (1993). A semiparametric bootstrap technique for simulating extreme order statistics. *J. Amer. Statist. Assoc.* **88**, 422, 477–485.

Zempléni, A. (1991). Goodness-of-fit for generalized extreme value distributions, *Technical Report*, University of Sheffield.

Index on Applications

Subject Index